FOCUS ON URBANIZATION TRENDS

FOCUS ON URBANIZATION TRENDS

LAURA M. DE SMET
EDITOR

Nova Science Publishers, Inc.
New York

For permission to use material from this book please contact us:
Telephone 631-231-7269; Fax 631-231-8175
Web Site: http://www.novapublishers.com

LIBRARY OF CONGRESS CATALOGING-IN-PUBLICATION DATA

Focus on urbanization trends / Laura M. De Smet (editor).
p. cm.
ISBN 978-1-60456-109-8 (hardcover)
1. Urbanization--Case studies. 2. Sustainable development--Case studies. 3. Urban ecology--Case studies. I. De Smet, Laura M.
HT361.F63 2008
307.76--dc22

2007051499

Published by Nova Science Publishers, Inc. ✦ New York

CONTENTS

Preface vii

Expert Commentary 1

An Analysis and System Construction on the Land Use of City Wetland Park 3
A-pei Zhao and Dong-qian Xue

Short Communications 15

Application of Computational Fluid Dynamics on the Prediction of Urban Thermal Environment and Urban Safety 17
Hong Huang, Ryozo Ooka and Shinsuke Kato

Urbanization: 21st Century Health Issues and Challenges in the North West province, South Africa 35
J.M. van Rooyen

Research and Review Studies 45

Chapter 1 Urbanization and the Increasing Risk from Mosquito-Borne Diseases: Linking Human Well-Being with Ecosystem Health 47
Paul T. Leisnham and David P. Slaney

Chapter 2 Urbanization, Economic Development and Income Inequality: A Demographic Perspective 83
Yasuhiro Sato and Kazuhiro Yamamoto

Chapter 3 Amyotrophic Lateral Sclerosis and Urbanization in the Health District of Ferrara, Italy, in 1964-1998 105
Vittorio Govoni, Enrico Granieri, Ilaria Casetta and Elisa Fallica

Chapter 4 China's Urbanization: New Trends and Challenges of Conceptualization 123
Li Zhang

Chapter 5 Biodiversity and Ecology of Hydrobionts in Natural and Anthropogenic Landscapes **143**
I.M. Khokhutkin, L.A. Kovalchuk and L.V. Chernaya

Chapter 6 Sustainable Rates of Groundwater Extraction in an Urbanized Plain, Central Japan **161**
Adrian H. Gallardo, Atsunao Marui, Shinji Takeda and Fumio Okuda

Chapter 7 Spatial Effects of the Interaction between Environment and Urban Area and Approaches to Coordination in Xi'an **177**
Dongqian Xue

Chapter 8 Measuring Urbanization in Low Income Countries for Public Health Interventions **191**
Andrew J. Tatem

Index **209**

PREFACE

Urbanization refers to a process in which an increasing proportion of an entire population lives in cities and the suburbs of cities. Historically, it has been closely connected with industrialization. When more and more inanimate sources of energy were used to enhance human productivity (industrialization), surpluses increased in both agriculture and industry. Larger and larger proportions of a population could live in cities. Economic forces were such that cities became the ideal places to locate factories and their workers. As the 21st century heats up the rate, pace, and complications posed by urbanization continue unabated. This new book presents the latest research in the field.

Based on the national land classification system and urban land classification system, the land use of city parks is analyzed and city park land classification system is constructed, as explained in the Expert Commentary. On the investigation, the land use classification of city wetland parks is divided. By the division and analysis, it represents the relations between land use and tourism development projects. It can standardize park land use administration and direct reasonable land use.

In recent years, the urban thermal environment and urban safety become highly concerned points for sustainable urban design. In the first Short Communication, in order to study the urban thermal environment and urban safety, Computation Fluid Dynamics (CFD) modeling has been applied. At first, a numerical simulation program which can be adapted to complex urban areas coupled with convection, radiation and conduction is developed and used to predict the urban thermal environment. The wind velocity, temperature, and humidity are obtained from the simulation, which shows good agreement with the results of the field measurement. Secondly, CFD model, turbulent combustion model, radiation model, and firebrand scattering model are coupled to predict the urban fire and the scattering of firebrands in a modeled urban area. It is found that when the inflow wind velocity is comparatively slow, the size of the thermal plume becomes significant, and when the inflow wind velocity is high, the thermal plume is suppressed and greatly inclines to the leeward side, which increases the risk of the fire spread to neighboring buildings. Firebrands are blown up by the thermal plume and the scattering distances are influenced by the background flow field, the magnitude of thermal plume, and the Stokes diameters of firebrands.

People in South Africa especially the black population (Africans) are subjected to massive urbanization since the early 1990's. They migrate from the traditional rural settings where they live under the auspices of a tribe and tribal habits to the large peri-urban informal settlements near all the larger cities and towns in South Africa in order to seek other opportunities to supply in their daily subsistence.

As explained in the second Short Communication, this has an impact on their health and adds to the chronic diseases of lifestyle risk factors typical of the western lifestyle they have to adopt. They are eating more salt, saturated fats and refined sugars, although high in energy compared to the traditional rural diet. Their intake of fiber and potassium are lower because they are eating less fruit and vegetables than their rural counterparts. They are more physical inactive than their rural counterparts, they walk less because they are making use of public transport and spending more time watching television. Migration to an urban settlement is a stressful life event and this daily experienced stress of poor living conditions leading to more violence, drug and alcohol abuse which may contribute to the larger incidence of cardiovascular diseases amongst them. The poor diet most pregnant women follows are leading to stunted offspring. All this contribute to the higher risk for chronic diseases and diseases of lifestyle like hypertension, diabetes and stroke. Stunting in children may also lead to hypertension and obesity in later life. The prevalence of most of this mentioned diseases of lifestyle are much higher in the urbanized subjects compared to their rural counterparts.

The major health burdens in South Africa are: infectious diseases like HIV/AIDS and tuberculosis, diseases related to malnutrition and chronic diseases as a cause of urbanization like hypertension and stroke a forerunner of vascular disease. The challenges would be a complementary dual approach to promote a healthy lifestyle for the whole population and the early diagnosis of disease in individuals with the cost-effective management of the risk factors and disease.

As presented in Chapter 1, urbanization is considered one of the primary drivers of ecological change because it alters the basic physical properties of the environment, including the vegetation, hydrology, soil structure, and topography. These changes in turn alter the complex abiotic (physical and chemical) and biotic processes that typified the original ecosystem. This often leads to the elimination or severe disruption of the ecological niches of resident species, whilst new ecological niches may be formed. The world is experiencing resurgence in infectious diseases attributable to urbanization and associated environmental changes, marking a third epidemiological transition. The most pre-eminent vector borne diseases to re-emerge and coincide with urbanization have been those vectored by mosquitoes. Urbanization has frequently increased mosquito-borne disease risk directly by providing dense numbers of susceptible human hosts, encroaching on areas of high mosquito density, and creating favorable adult and larval habitats. Urbanization has also increased mosquito-borne disease risk indirectly by causing changes (usually reductions) to the resident biodiversity that facilitate the increase and spread of native and invasive species. Today, New Zealand provides a case study nation where urbanization may facilitate an economic and public health catastrophe from mosquito-borne human disease. Despite no historical precedent, recent studies suggest that urbanization is increasing the risk of mosquito-borne human disease in New Zealand. Urbanization appears to be fostering the spread of resident mosquito species and increasing the likelihood of invasion from exotic vectors through its synergism with other anthropogenic changes, including climate change and increased international travel and trade, as well as natural factors, such as a depauperate native invertebrate fauna. Studying invasion processes help us understand how ecological shifts accompanying urbanization can increase the risk from disease vectors. This in turn will improve the surveillance and control of current exotic species, identify future successful invaders, and, ultimately, help limit the spread of mosquito-borne diseases. The proliferation of mosquito-borne diseases by urbanization highlights the intrinsic two-way link between ecosystem and human health. While traditional disease models can cope with the direct effect of an urban-mediated exposure for a specific disease outcome, the urbanization-mosquito relationship demands a multidisciplinary approach to understand the complex indirect ecological mechanisms leading to human disease.

Chapter 2 presents a model of endogenous fertility having two sectors (urban and rural) and analyzes the relationship among urbanization, fertility rate determination, and economic development. In the model, technological progress and human capital accumulation are complementary. The urban sector is assumed to have better opportunities for education and a more human capital-intensive production technology than the rural sector. A poverty trap is shown to exist, in which the economy has no human capital accumulation and no technological progress. Once the economy takes off from the poverty trap, human capital accumulation and technological progress advance, and the economy reaches an endogenous growth path. In order for the economy to take off from the poverty trap and begin economic development, sufficiently high human capital and technology levels are shown to be necessary. In addition, the model can replicate one observed pattern of economic development: with economic development, urbanization progresses, fertility declines, and income inequality first rises and then falls.

Chapter 3 develops the issue of the supposed relationship between the incidence of ALS and environmental factors, and in particular the supposed relationship with those environmental factors that are in relation with the level of urbanization or in relation with the rural environment. The hypothesis that changing in the frequency of the disease could be the consequence of environmental changes is supported. This chapter has first considered the available data from the international literature and has been more dedicated to the studies we have carried out in a well defined population and area since the 1960's that is the resident population of the Health District of Ferrara, Northern Italy. The analysis of the findings of these studies we have carried out using a better and more proper definition of rural and urban environment and the different urban environment levels in the study area reveals some interesting findings that have induced us to modify our previous ideas about this issue providing some interesting clues that seem very important for the future studies on ALS and environmental factors which should consider the main actual change in the environment that is the urbanization and the progressive depopulation of the rural area.

The unprecedented and intricate post-socialist urbanization process sets China apart not only from most countries but also from its early socialist state. As a result, many issues generated by the specific trajectory of China's urbanization remain elusive and present challenging opportunities for fruitful theoretical construction (Lin and Wei, 2002). By synthesizing and scrutinizing the latest official data, this paper examines recent changes in three spatial facets of China's urbanization, namely urbanization pace, rural-to-urban migration, and spatial reorganization of the urban system. On that basis, Chapter 4 identifies several issues which are important but remain under-discussion for theorization of China's post-socialist urbanization. The paper argues that to conceptualize Chinese post-socialist urbanization process requires, first, to explicate extant cutting-edge empirical studies and, then, to integrate their key propositions with a system-centered analytical framework.

Usually only terrestrial habitats are considered in the literature concerning investigations of the urban fauna ecology, while freshwater ecosystems are omitted. Leaches, benthic invertebrates, play a very important part in water ecosystems of various types. They are an important link in trophic chains: predators and blood-suckers for various groups of hydrobionts they are a prey for many fish species, desmans, otters, minks, etc. Some predatory leach species play a "sanitary" role in reservoirs, as they prey on carrion. Besides, they are additional and intermediate hosts for many parasites. The aim of Chapter 5 is to investigate the hirudofauna of reservoirs of various types (rivers, lakes, ponds, water reservoirs) in the South and Middle Urals. We compared the species composition and ecologo-physiological characteristics of leaches from 28 reservoirs located in urbanized and natural areas. We revealed the effect of anthropic factors on the qualitative and quantitative compositions of the leach fauna in the urban reservoirs.

Increased concentrations of pollutants (heavy metals) are marked in animal has tissues against the background of a decreased number of species and their abundance. This resulted from changes in the habitat and feeding conditions, although leaches were quite resistant to chemical agents. In the ponds of urban landscapes we marked the deficit of higher vegetation – substratum for leaches and for various hydrobionts groups, their food base.

Land subsidence due to excessive extraction of groundwater is a major environmental problem that has affected for several decades both the urban centers and the surrounding peri-urban farmlands scattered within the Nobi Plain, Central Japan. Even though strict regulations successfully reduced the subsidence rates, the threat is still far from being eliminated. The plain is one of the most important economic centers of Japan, sustaining an intensive agricultural and industrial production. Furthermore, its population density is among the highest in the country, averaging more than 1,000 persons per km^2. The recent economical growth of the region is expected to be accompanied by new developments and a higher demand for water supply therefore, authorities are compelled to update management strategies to satisfy the increasing demand without causing undesirable effects on the environment. Knowledge of the groundwater availability becomes an essential requisite for the establishment of these policies. A numerical simulation was applied to understand the groundwater flow in the region. The resulting model was then used to determine the safe yield of the aquifers under different management options. The analysis concentrated on the Aburashima site, as this area faces the possibility of development in the forthcoming years. Calculations indicated that withdrawal from the shallow aquifer is limited. In addition, a decline in piezometric levels reduced the groundwater availability in summer. Spreading a number of wells around the site constitutes an alternative to increase the pumping volume. Higher quantities and lower extraction costs makes the lower aquifer a more reliable source of groundwater. Chapter 6 constitutes one of the first steps towards the implementation of an integrated management plan of the water resources in the Nobi Plain, and constitutes a valid reference to update the current legislation regarding groundwater use in the region.

In Chapter 7, through analyzing the interaction between urbanization and environment, we sum up the spatial morphology, formation mechanism and types of the interaction between urbanization and environment. Then we generalize the spatial effect types of the interaction, namely tree-ring effect, decay effect and encroachment effect. In the case study of Xi'an City, we find out the spatial effects of the interaction between urbanization and environment and their spatial representation styles. Finally, we point out the macroscopic, mesoscopic and microscopic coordination arrangements, and our thoughts on the coordination in development mechanism, development structure, development ways and development goals.

Within the next 30 years, the global proportion of urban dwellers will rise from under half to two thirds of the world's population. Such a shift will entail massive public health consequences, and most of this urban transition will occur in low-income regions of the world. Urban populations face very different health risks compared to those in rural areas and, therefore, to target effective and relevant public health interventions, the need for clear, consistent data on urbanization pattern and extent is paramount. Decision makers are increasingly seeing satellite remote sensing and geographical information systems (GIS) as cost-effective solutions to monitoring urbanization at a range of spatial scales. Chapter 8 focuses on the progress made within the fields of remote sensing and GIS in terms of mapping, monitoring, and modeling urban environments for the effective targeting of limited public health resources across the developing world. Existing challenges, drawbacks, and future prospects are all discussed, along with specific examples that demonstrate how the creation of accurate urbanization maps aid the refinement of malaria transmission and outcome estimates across Africa.

EXPERT COMMENTARY

In: Focus on Urbanization Trends
Editor: Laura M. De Smet
ISBN: 978-1-60456-109-8

AN ANALYSIS AND SYSTEM CONSTRUCTION OF THE LAND USE OF A CITY WETLAND PARK

A-pei Zhao and Dong-qian Xue
College of Tourism and Environment,
Shaanxi Normal University, Xi'an 710062,China

Abstract

Based on the national land classification system and urban land classification system, the land use of city park is analysed, and a city park land classification system is constructed. In the investigation, the land use classification of a city wetland park is divided. By the division and analysis, it represents the relations between land use and tourism development projects. It can standardize park land use administration and direct reasonable land use.

Keywords: land use types, land use analysis, city wetland park

1. Question

Land use type has been researched for more than half a century throughout the world [1]. Since the 1930s, geographists from other nations have pointed to the idea of land use. Back in 1931, famous Soviet geographer J.I.C. Berg, in the Soviet landscape geographical zone, conceived the definition of the natural landscape and cited examples, writing that the natural landscape exists where terrain, climate, vegetation, and soil characteristics merge into a unified, harmonious whole and emerge in a certain zone again and again. American scholars J.O. Micro-qi (Veatch) defined a more exact concept of a natural type of land, building on the earlier concept of land types (1937). Records about land-use type in China date back to the early fifth century B.C. At the end of the 1950s, Zhao Song, Lin Chao and Chen Chuan-Kang surveyed and mapped land use type in the mountainous areas of Beijing, Chuanximaerkang and so on [1].

As for the existing land classification system, the National Land Classification System and Urban Land Classification systems have matured. However, there is a lack of data on park land use type, making it difficult for the authors to complete their work. Consequently, the authors began to collect the information to construct a suitable park land use classification system on the basis of theoretical and practical experience about the original land classification system, in hope that these can provide a useful reference for future park land analysis and land management.

2. The Theoretical Basis and Data Sources of the Land Classification

2.1. The Reference Source of Land Classification

The park land classification is based on the National Land Classification and Urban Land Classification, which is also combined with the park's features. The main reference documents come from the relevant files and regulations of the Ministry of Land and Resources.

2.2. The Theoretical Basis of Park Land Classification

The land classification entails merging each individual unit of land according to the quality's commonality or similarity through abstract generalization and combination at varying degrees, to obtain the different ranks of classified units.

2.2.1. The Principle of Park Land's Separation

(1)The General Principles

The land is a natural complex, which constitutes various elements of the land through their interaction and inter-restraint. Classification should based on the similarities and differences of the land; a comprehensive analysis of all elements of the land and a full consideration are required.

(2) The Principles of Dominant Factors

Regarding a comprehensive analysis of all elements of the land as a premise, the dominant factors of the leading role in certain conditions should be considered, so as to distinguish the similarities and differences of the land.

(3) The Practical Principles

The aim of land type study is the practical application of land management and providing future reference for the land use planning, so land classification should follow the principle of practicality.

2.2.2. The Land Classification Steps

(1) Identify the Classification Objects

It mainly depends on the task of studying and the scale of land type map to determine which level of land unit is regarded as a classified object module. Our study group analyses the city wetland park's land use, which aims to develop and utilize the wetland resources. Using 1:1000 large-scale maps, therefore, land points have been chosen as the main object.

(2) Identify the Basis of Classification and the Classification Indexes

The land use is relatively complex in the planning region, containing an assortment of water, land, and construction land. Therefore, when doing the land classification, factors such as hydrology, soil, vegetation and others must be fully considered at the same time.

2.2.3. The Methods of Land Classification

Due to the relatively small size of the planning region, the land classification used the sequence method.

3. System Construction of Park Used-Land Types

So far, the system of park used-land types has no references. Based on the former system of urban and town land use types and the entire system of national land classification, at last—through site survey, inspection and deliberation with the experts—this paper divides land use types of the city wetland park and works out the land classification system of the city wetland park.

3.1. The System of Urban and Town Used-Land Types

Mainly according to the administrative characteristics and use type, the classification program of urban and town land use in an urban and town land ownership survey regulation brought forward by the Land Administration Bureau, P.R.C., in September, 1989, divides urban and town land use types into 10 first-class types and 24 second-class types. This mainly applies to the classification of the current land use situation in established urban and town areas and residential areas.

Table 1. Urban and town land classification

First-class land types	Second-class land types
1. commerce and intellectual land	11. service industry land 12. tourism industry land 13. finance and insurance land
2. industrial warehouse land	21. industry land 22. warehouse land
3. municipal land	31. municipal public facilities land 32. green land
4. public building land	41. culture, sports and entertainment land 42. organ and propagation used-land 43. scientific research and design used-land 44. education used-land 45. health and medical used-land
5. dwelling land	61. railway land 62. civil airport 63. port and dock land 64. other transportation land

Table 1. (Continued)

First-class land types	Second-class land types
6. transportation land	
7. special land	71. military facilities land 72. concering foreign land 73. religion land 74. jail land
8. water area land	
9. farm-oriented land	91. paddy land 92. vegetable land 93. dry land 94. garden land
10. other land	

Notice: Evaluate teacher qualification examination committee, Ministry of Land and Resources land edit. Land management basis (2002).

3.2. The National Land Classification System

The Ministry of Land and Resources P.R.C. tried to put a new land use classification system (land classification) into effect beginning January 1, 2002.

Table 2. Test of land classification in 2001

First-class land types	Second-class land types	Third-class land types (71 types)
1.agricultural land	11. cultivated land	111. irrigated paddy fields 112. rain fed paddy fields 113 irrigated paddy land 114. dry land 115. vegetable land
	12. garden plots	121. orchards 122. mulberry fields 123. tea plantations 124. rubber plantations 125. other plantations
	13. wood land	131. forest land 132. shrubbery land 133. few trees land 134. man-made forest land 135. mark land 136. seeding nurseries land etc.
	14. grass land	141. natural grass land 142. improved grass land 143. man-made grass land
	15. other agricultural land	151. livestock and poultry breeding land 152. agriculture establishment land 153. path 154. pond surface 155. fishery water surface 156. farm hydraulic structures 157. ridge 158. grain sunning ground etc.

Table 2. Continued

First-class land types	Second-class land types	Third-class land types (71 types)
2. building land	21. commercial and service land	211. commercial land 212. finance and insurance land 213. hotel land 214. other land
	22. industry and storage land	221. industry land 222. mining land 223. warehouse land
	23. public facility land	231. basic public project land 232. landscape recreation land
	24. public building land	241. government land 242. education land 243. research land 244. sport and culture land 245. medical and healthy land 246. welfare land
	25. dwelling land	251. urban single housing land 252. combined housing land 253. rural housing base land 254. unused housing base land
	26. transportation land	261. railway land 262. highway land 263. civil airport land 264. harbors and wharfs 265. pipeline land
	27. hydraulic establish land	271. reservoir surface 272. water conservancy land
	28. special land	281. military land 282. embassy land 283. religion land 284. prison land 285. funeral land
3. unused land	31. unused land	311. waste grass land 312. saline-alkali land 313. wetland 314. sand land 315. bare land 316. exposed rock and shingle land 317. other unused land
	32. other land	321. river surface 322. lake surface 323. reed land 324. beaches and flats 325. glaciers and firns.

Notice: According to a document provided by The Ministry of Land and Resources P.R.C.

3.3. The System Construction on the Land Use of Wetland Park

With the national land classification system and the town land category system as a foundation, according to the land of use, conduct principle, use and overlay character, the land use classification has been done as shown in Table 3, which takes sustainable use, comprehensive and usability for human needs and unitarity as a principle.

The present land use condition varies by area, so adjustments must be made in land use analysis.

Table 3. Park land classification

First-class land types	Second-class land types	Third-class land types (71 types)
1. green land	11. garden land	111. fruit ranch 112. flower garden 113. other garden land
	12. woodland	121. forest land 122. shrubbery land 123. few trees land 124. man-made forest land 125. mark land 126. seeding nurseries land etc.
	13. grass land	131. natural grass land 132. improved grass land 133. man-made grass land
	14. other green land	141. dish garden 142. facility green land 143. green path
2. building land	21. landscape building land	
	22. dwelling land	
	23. service and manage land	231. service land 232. manage land etc.
	24. transportation land	241. road land 242. park land 243. service facility land etc.
	25. hydraulic project land	
	26. public sanitization land	
	27. other building land	
3. water suface		
4. unused land		

4. Case Research and Analysis of Land Use

4.1. System Construction on the Land Use of Xi'an Baliu City Wetland Park

Above the research and actual circumstance of Baliu City Wetland Park, a land type system for Baliu City Wetland Park is constructed. Xi'an Baliu City Wetland Park includes both sides of the large dike of the Ba River; there is a great difference between the land use of the two sides which is very difficult to reconcile. Therefore, the system of land use has three parts: inside the dike, on the dike and outside the dike. The part inside the dike maintains its original natural wet state; the part on the dike is used for transportation; the park outside the dike is more complicated, and its land use types are more abundant. The details are shown in Table 4.

Table 4. Park land classification

	First-class land types	Second-class land types	Remarks
inside the dike	water surface island washland		
on the dike			transportation
outside the dike	eco-green space	woodland grass land mixed land including woodland and grass land	
	garden-plot	orchard vegetable	
	water land	lake land water channel eyot	
	landscape architecture		benchmark architecture and building pieces
	housing land		residence for farmers such as Lanjia village
	transportation land	road land park land service facility land	
	service land	hotel land shopping land	
	land for tourism activity		land for tourism facilities, e.g.
	management land		
	hydraulic project land		30 meters on each side of the dike
	public sanitization land		W.C. and wastebin
	other land		

4.2. Analysis of Land Use for Baliu City Wetland Park

4.2.1. The Data Source and Technique Support

After creating the area topology in the planning map which used software (include ArcGIS, ArcView), classification and stat., the original data was gathered. Based on the original data and using Excel to calculate and analyse, the last data was obtained.

4.2.2. Itemized Table of Xi'an Baliu Wetland Park Land Classification

Table 5. Itemized table of park land classification

	Area (km^2)	Ratio of the whole park area (%)	First-class land types	Area (km^2)	Ratio of the whole park area (%)
inside the dike	1.91	39.96	island	0.440	9.21
			washland	0.850	17.78
			water surface	0.620	12.97
on the dike	0.03	0.63			
outside the dike	2.84	59.41	eco-green land	1.210	25.31
			garden land	0.260	5.44
			water surface	1.000	20.92
			landscape architecture	0.001	0.02
			transportation land	0.100	2.09
			service land	0.007	0.15
			housing land	0.100	2.09
			tour activity land	0.005	0.10
			management land	0.002	0.04
			hydraulic project land	0.070	1.46
			other land	0.085	1.79
sum	4.78	100			

The analysis for transportation land and building land is as follows:

(1) Analysis for Transportation Land

Transportation land includes path land, park land and relative service land. The classification in detail is given in Table 6.

Table 6. Itemized table of the transportation land classification

No.	Land type	Area (m^2)	Ratio of the whole park area (%)
1	path land	84262.3	1.76
2	park land	11225.3	0.23
3	dock land	350	0.01
4	total	96837.6	2.00

(2) Analysis for Green Land

Green land for the wetland park includes forest land, grassland and a mixed area of forest and grass land. The classification in detail is listed in Table 7.

Table 7. Itemized table of green land classification

No.	Land type	Area (m^2)	Ratio of the whole park area (%)
1	forest land	328,921	6.88
2	grass land	480,000	10.42
3	mixture land of forest and grass land	404,968	8.47
4	total	1,213,889	25.77

4.2.3. The Analysis for Land Use—Comprehensive Quantity Characteristic of Structure

The land use structure in a district can use diversity, dominance and fragmentation as an index to analyse.

(1) Diversity: comprehensive description for land-use abundant degree and even degree, the calculation formula is [3-5]:

$$H= \sum_{i=1}^{S} Pi \ln Pi \quad (1)$$

in the formula, "S" is the number of land use type; "Pi" is the ratio of land tape *i* in the whole area; "H" is diversity. The greater "H" is, the more abundant is the eco-landscape.

(2) Dominance: Used to measure the dominant degree of one land use type or a few land use types in the land use structure, the calculation formula is [3-5]:

$$D=\ln S+\sum_{i=1}^{S} Pi \ln Pi \quad (2)$$

in the formula, "S" is the number of land use type; "D" is dominance. The bigger "D" is, the stronger is the dominance by one or a few land use type in the land use structure.

(3) Fragmentation: the spot number per square kilometer. It means the broken degree of spot pieces in land; the calculation formula is [3-5]:

$$F=\sum_{k=1}^{S} Nk / A \quad (3)$$

in the formula, "S" is the number of land use type; "N_k" is the spot number for the land use type K; "A" is the whole area; "F" is fragmentation. The bigger "F" is, the greater is the land's fragmentation.

For disclosing the character of park land use structure, the writer uses the three formulas above to compute the diversity, dominance and fragmentation for the fourth district in the Xi'an Baliu City Wetland Park. The four districts are: key protected area, wetland exploded area, tourist activity area and management and servicing area.

Table 8. Index of diversity, dominance and fragmentation of land use in different zones

Subarea	Key protected area	Wetland exploded area	Tourist activity area	Management and servicing area
Diversity	0.79164	0.9712	1.3842	0.7594
Dominance	0.3070	0.1274	0.5617	0.6269
Fragmentation	6.7010	15.8730	50.2994	18.5185

4.2.4. Structure Analysis of the Land Use of City Park

Based on the studies above, it can be seen clearly that the main land use types of Xi'an BaLiu City wetland Park are water surface and eco-green land; the two modes occupy 20.92% and 25.31% of the park's total area individually, and the two together occupy more than three quarters of the external bank area, nearly half of the park's total area. The planning ideals of "ecology and nature" were strictly held so as to construct good conditions for the wetland ecosystem. The construction land use accounts for a very small proportion: the transportation facility, the servicing facility, the management facility as well as the landscape construction land use all together account for only 2.3% of the park's total area; the inhabitant land use accounts for 2.09% of the park's total area, and the sum of the two does not reach 5% the park's total area.

An analysis from the point of view of multiple utilization of the land areas should focus on the following: the tour activity area, the wetland exploded area, the key protected area, and the management and servicing area. Obviously, the structure of the tour area land utilization is relatively balanced. The structures of the key protected area and the management-servicing area land utilization are relatively simple, which is also manifested in designing different traveling activity projects in different regions: the park's tour projects mostly concentrate in the tour area; therefore, this area's land utilization types are relatively greater, while other regions' traveling functions are relatively fewer, and the land utilization type is relatively unitary. On the other hand, different regions' functions are different and the land utilization structures are not the same: the key protected area is the wetland's core region, and the original condition has been basically preserved, so the land use structure is unitary. On the other hand, the tour area is the primary sightseeing and leisure area for tourists and the land utilization structure is rich.

In analyzing the land utilization's priorities, the four regions' differences are not great: water areas and the ecological green area all maintain absolute superiority in these four regions. Obviously, it can be seen from the computed results of land utilization the degree of fragmentation. The values of the tour area are greater than other regions, once more confirming its relationship with tourist activity projects: ecological roving zones like the aquatic labyrinth have been designed, the land mottling is considerable, and the degree of fragmentation is high.

Conclusion and Discussion

The impact of the city park on an urban population's city life is greater and greater, its planning and development affecting the ecological environment of a peripheral locality as it maintains a close correlation with economic development (real estate in particular). Therefore, the city park's land utilization and management appears especially important. In foundation of the original land classification, this article unified the characteristics of city park land use, constructed the system type of city park land use, took the planning Baliu City Wetland Park as an example to conduct the case research, simultaneously carried out the land utilization analysis, and drew the following conclusions:

(1) The overall structure of park land utilization is related to the park's subject.
(2) The rich degree of park land utilization structure is related to the park's design of tourist projects: the more projects, the more complex the land utilization; otherwise, the more simple.
(3) In different regions, as the functions are different, the land utilization structures are also different. Regarding to the protected regions, the land utilization structures are unitary; in the open regions, the land utilization structures are rich.

City parks with different subjects can have different land utilization emphases; therefore, in concrete practical application, we should make the adjustment to the land utilization type system, unifying the actual situation slightly in order to better service the land statistics, management and planning.

References

[1] Ni Zhaoxiang. 1999.6. *An introduction to Land Classification and Evaluation.* Higher Education Press.
[2] Liu Pinghui, Guo jinmin. 2005.3. *Land Use Classification and the Developing and Planning of Urban and Rural.* Chinese Earth Publisher.
[3] Tourner M R H Gadner (eds). *Quantitative methods in landscape ecology* [M]. New York: Springer-Verlag, 1991.
[4] Yue Wenze,Xu Jianhua,Jin Jiong, etc. 2002. Study on Land-use Structure of Suburb and Fractal Model of Semiarid Area in Northwest China: A Case of Xigu District in Lanzhou City. *Journal of Desert Research* **8**(22):249-255.
[5] Zhang Jintun, Qiu Yang, Zheng Fengying, etc. 2000. Quantitative Methods in Landscape Pattern Analysis. *Journal of Mountain Research* **4**(18):346-352.
[6] Xiao Duning. *Land Ecology: Theory Method and Application* [M]. Beijing: Chinese Forestry Publisher.

SHORT COMMUNICATIONS

In: Focus on Urbanization Trends
Editor: Laura M. De Smet
ISBN: 978-1-60456-109-8

APPLICATION OF COMPUTATIONAL FLUID DYNAMICS ON THE PREDICTION OF URBAN THERMAL ENVIRONMENT AND URBAN SAFETY

Hong Huang**, *Ryozo Ooka and Shinsuke Kato
Institute of Industrial Science, The University of Tokyo,
4-6-1 Komaba, Meguro-ku, Tokyo, 153-8505, Japan

Abstract

In recent years, the urban thermal environment and urban safety become highly concerned points for sustainable urban design. In this study, in order to study the urban thermal environment and urban safety, Computation Fluid Dynamics (CFD) modeling has been applied. At first, a numerical simulation program which can be adapted to complex urban areas coupled with convection, radiation and conduction is developed and used to predict the urban thermal environment. The wind velocity, temperature, and humidity are obtained from the simulation, which shows good agreement with the results of the field measurement. Secondly, CFD model, turbulent combustion model, radiation model and firebrand scattering model are coupled to predict the urban fire and the scattering of firebrands in a modeled urban area. It is found that when the inflow wind velocity is comparatively slow, the size of the thermal plume becomes significant, and when the inflow wind velocity is high, the thermal plume is suppressed and greatly inclines to the leeward side, which increases the risk of the fire spread to neighboring buildings. Firebrands are blown up by the thermal plume and the scattering distances are influenced by the background flow field, the magnitude of thermal plume, and the Stokes diameters of firebrands.

1. Introduction

Urban problems, such as, urban heat island, urban air pollution and urban safety, become more and more serious recently due to the progresses of mega cities. The revolutionary progress in Computational Fluid Dynamics (CFD) modeling makes it possible to model urban

* E-mail address: hhong@iis.u-tokyo.ac.jp

boundary layer problems. In this study, in order to study the urban thermal environment and urban safety, Computation Fluid Dynamics (CFD) modeling has been applied. At first, a numerical simulation program which can be adapted to complex urban areas coupled with convection, radiation and conduction is developed and used to predict the urban thermal environment (Huang et al., 2005). The wind velocity, temperature, and humidity are obtained from the simulation, which shows good agreement with the results of the field measurement. Secondly, CFD model, turbulent combustion model, radiation model and firebrand scattering model are coupled to predict the urban fire and the scattering of firebrands in a modeled urban area (Huang et al., 2006).

2. Application on Urban Thermal Environment

2.1. Urban Thermal Environment Simulation Method

The urban thermal environment numerical simulation is based on a simulation method coupled with three-dimensional CFD analysis, three-dimensional radiation analysis, and one-dimensional heat conduction analysis (Harayama et al., 2002; Yoshida et al., 2000a; Yoshida et al., 2000b; Chen et al., 2004; Huang et al., 2005). Figure 1 shows the flowchart for this numerical simulation. First, boundary conditions are set up from various input conditions. Second, a three-dimensional radiation calculation is performed. Then, the temperature distribution inside the ground or wall is calculated by solving an unsteady one-dimensional heat conduction equation. Three-dimensional coupled convection and water vapor transportation calculations are performed continuously by adding new boundary conditions for the surface temperature distribution of the ground and wall and air-conditioning heat load obtained from the radiation and conduction calculation. The coupled simulation of convection, radiation and conduction is then completed by repeating these operations in series. This numerical simulation adapts for an unstructured computational grid suitable for a complex urban area.

The CFD governing equations of continuity, momentum, temperature, humidity (pollutants), turbulent energy and turbulent dissipation rate are shown in Table 1. Turbulence is modeled using the standard k-ε closure. The heat balance at the outdoor surfaces of the ground and building walls are calculated by the heat balance equation

$$S_i + R_i + H_i + LE_i + C_i = 0$$

at each solid face i (Figure 2), and the surface temperature T_i is obtained.

On the left side of the equation, a positive value shows an inflow of energy to solid face i, and whereas a negative value shows an outflow. S_i, R_i are absorbed solar radiation and net long wave radiation respectively; H_i, LE_i, C_i are convective heat transfer at the solid surface, amount of heat dissipation by evaporation from the surface and amount of conduction heat to the building and ground. The detail of calculations for radiation and surface temperature can be found in the paper of Huang et al., (2005).

Table 1. Governing equations

Continuity equation

$$\frac{\partial\langle u_i\rangle}{\partial x_i}=0$$

Momentum equation

$$\frac{\partial\langle u_i\rangle}{\partial t}+\frac{\partial\langle u_i\rangle\langle u_j\rangle}{\partial x_j}=-\frac{1}{\rho}\frac{\partial\langle p\rangle}{\partial x_i}+\frac{\partial}{\partial x_j}\left(\nu\left(\frac{\partial\langle u_i\rangle}{\partial x_j}+\frac{\partial\langle u_j\rangle}{\partial x_i}\right)-\langle u_i'u_j'\rangle\right)-g_i\beta(\langle\theta\rangle-\theta_0)$$

Temperature transport equation

$$\frac{\partial\langle\theta\rangle}{\partial t}+\frac{\partial\langle u_i\rangle\langle\theta\rangle}{\partial x_i}=\frac{\partial}{\partial x_i}\left(\alpha\frac{\partial\langle\theta\rangle}{\partial x_i}-\langle u_i'\theta'\rangle\right)$$

Pollutant transport equation

$$\frac{\partial\langle C_i\rangle}{\partial t}+\frac{\partial\langle u_i\rangle\langle C_i\rangle}{\partial x_i}=\frac{\partial}{\partial x_i}\left(D\frac{\partial\langle C_i\rangle}{\partial x_i}-\langle u_i'C_i'\rangle\right)$$

Turbulent energy transport equation

$$\frac{\partial k}{\partial t}+\frac{\partial\langle u_j\rangle k}{\partial x_j}=P_k+G_k+D_k-\varepsilon$$

Turbulent dissipation rate equation

$$\frac{\partial\varepsilon}{\partial t}+\frac{\partial\langle u_j\rangle\varepsilon}{\partial x_j}=D_\varepsilon+\frac{\varepsilon}{k}\left(C_{\varepsilon 1}P_k+C_{\varepsilon 3}G_k-C_{\varepsilon 2}\varepsilon\right)$$

$$-\langle u_i'u_j'\rangle=\nu_t\left(\frac{\partial\langle u_i\rangle}{\partial x_j}+\frac{\partial\langle u_j\rangle}{\partial x_i}\right)-\frac{2}{3}k\delta_{ij} \qquad \nu_t=C_\mu\frac{k^2}{\varepsilon}$$

$$-\langle u_i'\theta'\rangle=\frac{\nu_t}{\sigma_\theta}\frac{\partial\langle\theta\rangle}{\partial x_i} \qquad -\langle u_i'C_i'\rangle=\frac{\nu_t}{\sigma_w}\frac{\partial\langle C_i\rangle}{\partial x_i}$$

$$P_k=-\langle u_i'u_j'\rangle\frac{\partial\langle u_i\rangle}{\partial x_j} \qquad G_k=-g_i\beta\langle u_i'\theta'\rangle$$

$$D_k=\frac{\partial}{\partial x_j}\left(\frac{\nu_t}{\sigma_k}\frac{\partial k}{\partial x_j}\right) \qquad D_\varepsilon=\frac{\partial}{\partial x_j}\left(\frac{\nu_t}{\sigma_\varepsilon}\frac{\partial\varepsilon}{\partial x_j}\right)$$

$\sigma_k=1.0,\quad \sigma_\varepsilon=1.3,\quad \sigma_\theta=0.9,\quad \sigma_w=0.9,$

$C_{\varepsilon 1}=1.44,\quad C_{\varepsilon 2}=1.92,\quad C_{\varepsilon 3}: 1.44(G_k>0);0(G_k<0)$

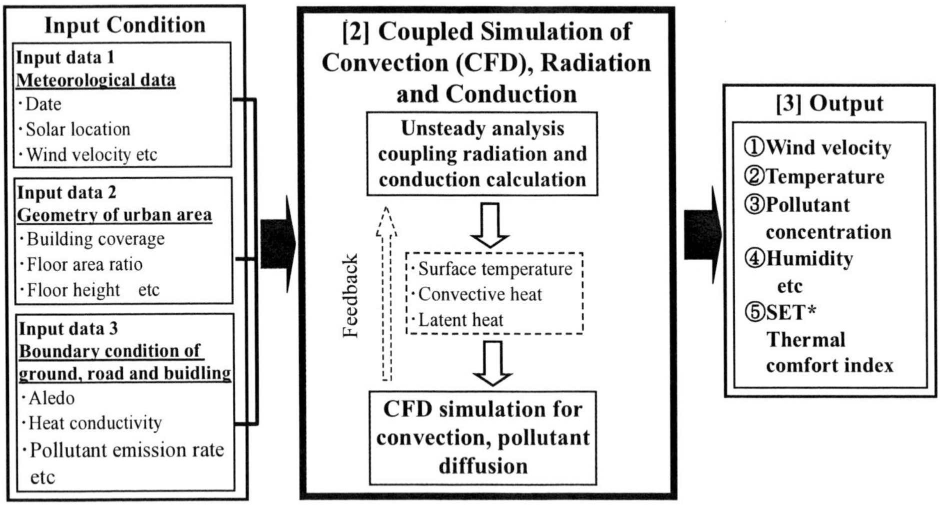

Figure 1. Flowchart of the numerical simulation.

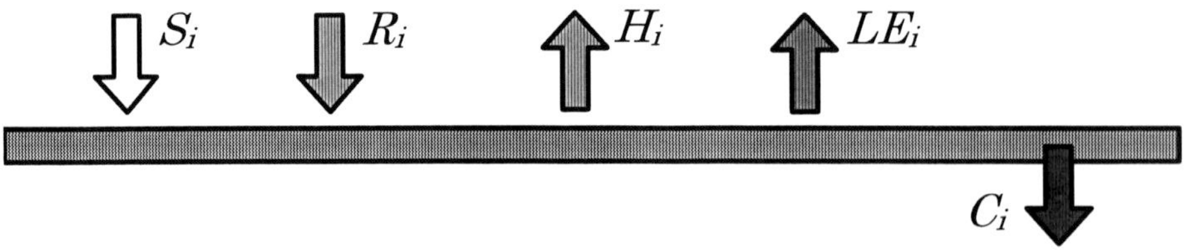

Figure 2. Heat balance of solid surface i.

2.2. Outline of Computation

The simulation object is shown in the frame enclosed by broken lines in Figure 3, where the Shinjuku (Tokyo, Japan) DHC (District Heating and Cooling Systems) center (next to Shinjuku Park Tower, a 52-storey building) is. Field measurements of outdoor thermal environment were taken here on August 20th-22nd, 2003 to investigate the influence of the heat release of the DHC center. An unstructured computational grid is created based on three-dimensional city model data. August 22nd at 13:00 is targeted as the date and time for the analysis. Total solar radiation is set at 701 W/m^2 based on the measurements. The building wall is assumed to be concrete here. The thickness of the wall is set to be 28 cm. Table 2 shows details of the analysis conditions. The inflow wind velocity is set at 1/4th power profile, and the measurement value for the Shinjuku Park Tower rooftop (2.0 m/s, southwest, and H = 220 m) is used as the representative velocity of the wind (U_0 in Table 2). The constant flux layer assumption is used to evaluate the turbulent kinetic energy.

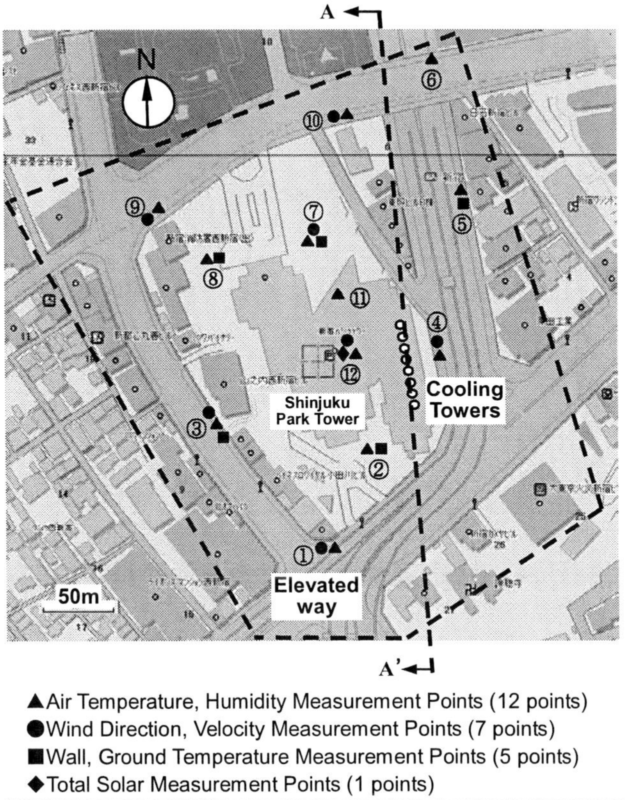

Figure 3. Measurement and analysis field.

Table 2. Analysis conditions

Turbulent model	Standard k-ε model (Inclusion of buoyancy effect)
Difference scheme	Convection terms: MARS*
Inlet	$U = U_0 \cdot (Z / Z_0)^{1/4}, Z_0 = 220m$ $k = 1.5 \cdot (I \times U_0)^2, I = 0.1$ $\varepsilon = C_\mu \cdot k^{3/2} / l$ $l = 4(C_\mu \cdot k)^{1/2} Z_0^{\ 1/4} Z^{3/4} / U_0$
Side, sky	Free slip
Wall	Generalized logarithmic law

*MARS: Monotone Advection and Reconstruction Scheme, hybrid second-order scheme (STAR-CD, 2001).

The turbulence scale is set based on the method devised by Murakami et al. (1988). The average values of each measurement point are used for the initial temperature and humidity conditions. The air-conditioning heat released from the DHC system is set to a design value of 200,000 kW, and the ratio of sensible heat to latent heat is set at 1:9. The air-conditioning heat released from buildings that are not supplied by the DHC system (named individual buildings here) is obtained by the radiation and conduction coupled calculation. The internal heat generation of building is set to 35W m^{-2}. The ratio of sensible heat to latent heat is set at 6:4 for these individual buildings. Moreover, the velocity of the heat released from the cooling tower of the DHC system is set at 16.75 m s^{-1} in an upward direction, which is the value at the maximum operating time.

2.3. Simulation Results and Comparison with Field Measurements

Figure 4 shows the simulation results for the horizontal distribution of the wind velocity and the measurement results at a height of 1.5 m. As an overall tendency, the simulation shows good agreement with the measurements. The velocity of the wind decreases close to the building, and a vortex is formed on the north side of the Shinjuku Park Tower (around measurement point 7). Moreover, the simulation values for the scalar wind velocity are 0.73 m/s at point 3, 0.98 m/s at point 4, and 0.89 m/s at point 9; while the measured results are 0.94 m/s, 1.09 m/s, and 0.84 m/s respectively, which show very good agreement. The wind direction is also well simulated at points 3, 4, 7 and 10. The vertical distribution in section A-A' (refer to Figure 3) is shown in Figure 5. It shows that a strong rising flow is formed by the influence of the upward exhaust from the cooling tower. It implies that the heat and vapor released from the DHC system are blown upwards into the sky.

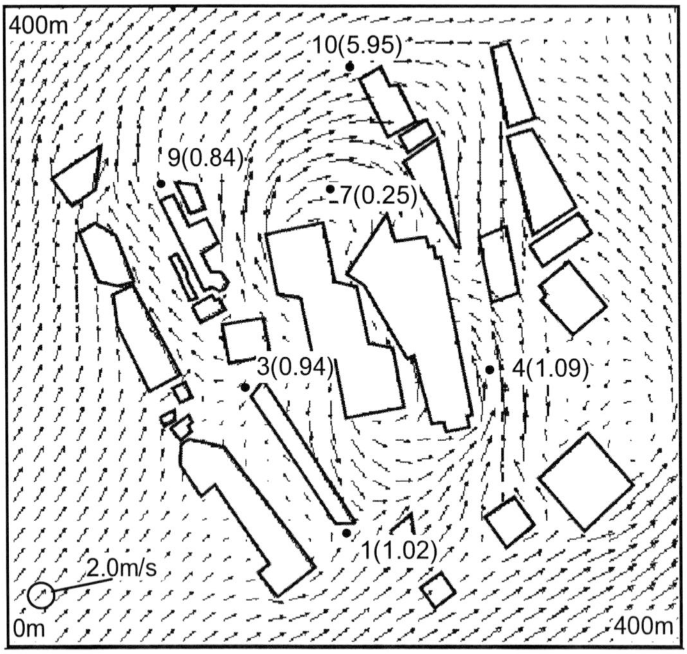

Figure 4. Horizontal wind distribution (m/s)(1.5m Height( ) measured wind velocity).

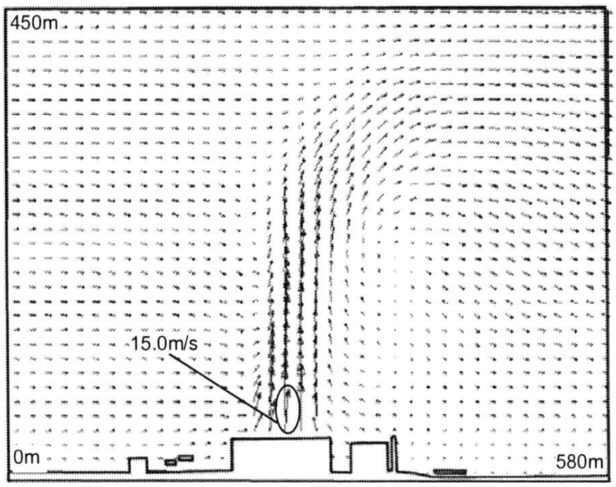

Figure 5. Vertical wind distribution (m/s) (A-A' section).

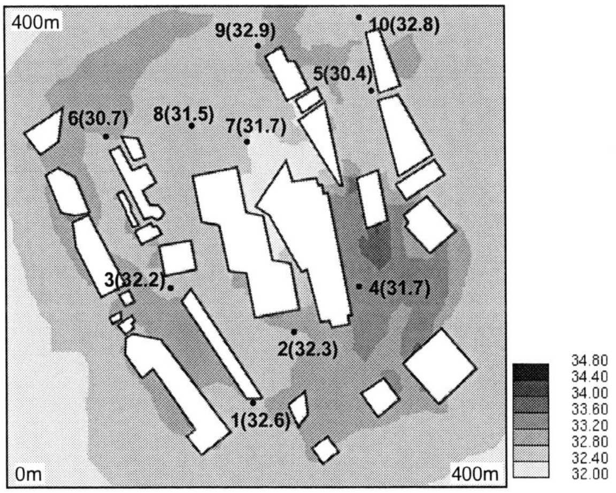

Figure 6. Horizontal temperature distribution (°C) ( ) measured temperature (1.5m Height).

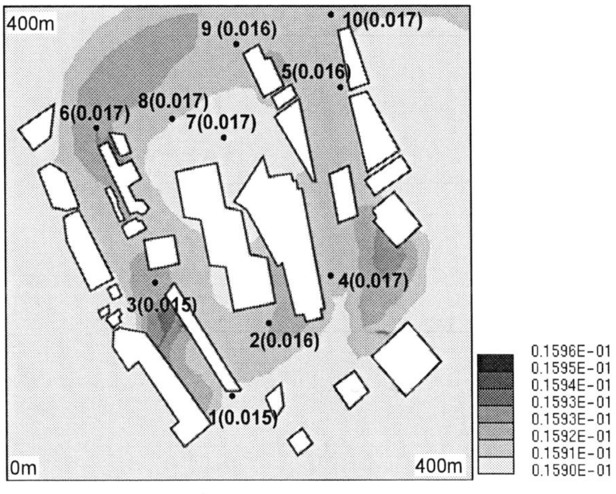

Figure 7. Horizontal absolute humidity distribution (kg kg'$^{-1}$) ( ) measured absolute humidity (1.5m Height).

The horizontal distributions of temperature and absolute humidity at a height of 1.5 m are shown in Figures 6 and 7. It can be seen that the temperatures are higher around measurement points close to buildings and places where the wind velocity is weak. We can see that the measured results and the simulation results are in good agreement. The vertical distribution at section A-A ' is shown in Figure 8 and 9. It can be seen that temperature stratification occurs around the buildings and the ground due to the heat flux from the buildings and the ground in the vertical direction. It is also thought that there is little influence on the thermal environment at the pedestrian level because the heat and vapor are expelled from the cooling tower high up into the sky. This agrees with the measured result that there is little difference between the temperature and humidity on the windward and leeward sides of the area.

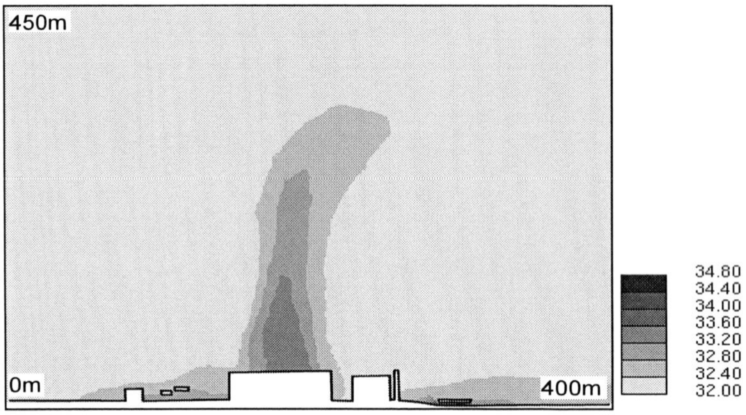

Figure 8. Vertical temperature distribution (°C) (A-A' section).

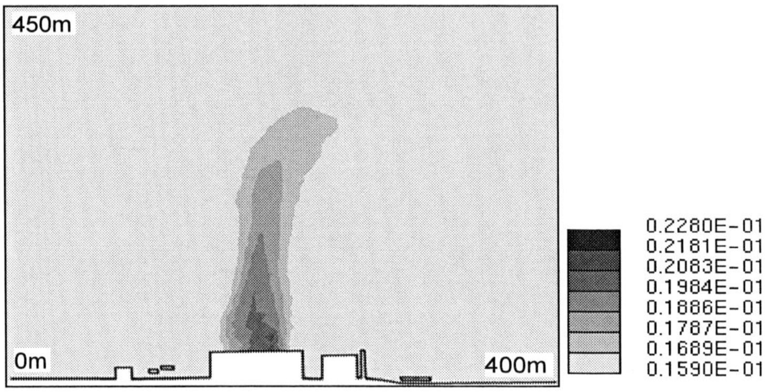

Figure 9. Vertical absolute humidity distribution (kg kg'$^{-1}$) (A-A' section).

3. Application on Urban Safety

The factors that can cause urban fire spread are contact fire, radiation, convection and firebrands. From investigations of past urban fires, fire spread by firebrands almost occurred, especially in the case of fires that cause extensive damage (Iwami et al, 2002; Ohmiya et al, 1999). However, there are few findings on the properties and mechanisms of firebrands because there are many uncertain factors, such as meteorological conditions, the combustion conditions in the building that is on fire, the types of building materials, and so on. In this study, in order to predict urban fire and firebrand scattering, CFD, turbulent combustion model, radiation model and firebrand scattering model are coupled. Then, the density, terminal velocity and Stokes diameters of firebrands which are necessary for the firebrand scattering model are measured. Finally, an urban fire simulation in a modeled urban area has been conducted using the coupled model and the firebrand experimental data. The firebrands scattering model used here focuses on the flying of firebrands, the combustion time of firebrands is thought to be much longer than the flying time. It is also assumed that the temperature of firebrands gives no influence on the air flow, and the mass and shape are not changed during flight.

3.1. Model Description

3.1.1. The CFD Model

Both the density and the temperature in the fire plume fluctuate widely. Therefore, the assumption of an incompressible fluid (Boussinesq approximation) does not hold, and the plume must be considered as a compressible fluid. Thus, the Favre-averaged process is introduced here for all the transport equations. Table 3 shows the governing equations. A modified k-ε model, as proposed by El Tahry (1983) for compressible reciprocating engine flows, is used. The main difference between the modified model and the standard model is the last term in the ε equation which results from the effects of compressibility.

A fire model is required to close the above governing equations. One of the common methods is to replace the fire region by a volumetric heat source (or sometimes heat flux). However, a combustion model that accounts for the chemical reactions between fuel and oxygen must be included to describe the actual fire more closely. Therefore, combustion in the gaseous phase is modeled using the eddy dissipation combustion model (Magnussen et al, 1976) here, which is widely used in fire modeling. The chemical reaction rate is proportional to the turbulence eddy decay rate and minimum concentration of the fuel or oxygen:

$$\overline{w}_f = -\overline{\rho}(\varepsilon / k)\min(A\widetilde{Y}_f, A\widetilde{Y}_{O_2} / i)$$

where A takes the value 4.0, and i is the stoichiometric ratio of oxygen to fuel. The following chemical reaction is assumed (Novozhilov et al, 1996):

$$C_nH_m + (n - s + \frac{m}{4})(O_2 + 3.76N_2) \rightarrow sC + (n - s)CO_2 + \frac{m}{2}H_2O + 3.76(n - s + \frac{m}{4})N_2$$

Here s is a parameter that defines the amount of soot produced. A zero value means complete combustion reaction is obtained to give carbon dioxide and water. When s is non-zero, some of the carbon remains as soot, with a consequent reaction to become CO_2 finally. The value of s can be decided from the soot conversion factor, which is chosen from experimental data; for example, 2% for propane (Yan et al, 1996). The soot concentration is determined from the chemical species equation.

Table 3. Governing equations

$$\frac{\partial \bar{\rho}}{\partial t}+\frac{\partial(\bar{\rho}\tilde{u}_j)}{\partial x_j}=0$$

$$\frac{\partial(\bar{\rho}\tilde{u}_i)}{\partial t}+\frac{\partial}{\partial x_j}(\bar{\rho}\tilde{u}_i\tilde{u}_j)=-\frac{\partial \bar{p}}{\partial x_i}+\frac{\partial}{\partial x_j}(\bar{\tau}_{ij}-\bar{\rho}\widetilde{u_i^{"}u_j^{"}})+\bar{\rho}g_i$$

$$\frac{\partial(\bar{\rho}\tilde{h})}{\partial t}+\frac{\partial}{\partial x_j}(\bar{\rho}\tilde{u}_j\tilde{h})=\frac{\partial \bar{p}}{\partial t}+\frac{\partial}{\partial x_j}(\frac{\lambda}{c_p}\frac{\partial \tilde{h}}{\partial x_j}-\bar{\rho}\widetilde{u_i^{"}h^{"}}+\sum_{n=1}^{N}\bar{h}_n\bar{\rho}D_n\frac{\partial \tilde{Y}_n}{\partial x_j})+\bar{u}_j\frac{\partial \bar{p}}{\partial x_j}+\bar{\tau}_{ij}\frac{\partial \bar{u}_i}{\partial x_j}-\bar{S}_R$$

$$\frac{\partial(\bar{\rho}\tilde{Y}_n)}{\partial t}+\frac{\partial}{\partial x_j}(\bar{\rho}\tilde{u}_j\tilde{Y}_n)=\frac{\partial}{\partial x_j}(\bar{\rho}D_n\frac{\partial \tilde{Y}_n}{\partial x_j}-\bar{\rho}\widetilde{u_i^{"}Y_n^{"}})+\bar{w}_n$$

$$\frac{\partial(\bar{\rho}\tilde{k})}{\partial t}+\frac{\partial(\bar{\rho}\tilde{u}_j\tilde{k})}{\partial x_j}=\frac{\partial}{\partial x_j}((\mu+\frac{\mu_t}{\sigma_k})\frac{\partial \tilde{k}}{\partial x_j})+\mu_t(\frac{\partial \tilde{u}_i}{\partial x_j}+\frac{\partial \tilde{u}_j}{\partial x_i})\frac{\partial \tilde{u}_i}{\partial x_j}-\frac{2}{3}(\mu_t\frac{\partial \tilde{u}_k}{\partial x_k}+\bar{\rho}\tilde{k})\frac{\partial \tilde{u}_i}{\partial x_i}-g_i\frac{\mu_t}{\sigma_h}\frac{1}{\bar{\rho}}\frac{\partial \bar{\rho}}{\partial x_i}-\bar{\rho}\tilde{\varepsilon}$$

$$\frac{\partial(\bar{\rho}\tilde{\varepsilon})}{\partial t}+\frac{\partial(\bar{\rho}\tilde{u}_j\tilde{\varepsilon})}{\partial x_j}=\frac{\partial}{\partial x_j}((\mu+\frac{\mu_t}{\sigma_\varepsilon})\frac{\partial \tilde{\varepsilon}}{\partial x_j})+$$

$$C_{\varepsilon 1}\frac{\tilde{\varepsilon}}{\tilde{k}}(\mu_t(\frac{\partial \tilde{u}_i}{\partial x_j}+\frac{\partial \tilde{u}_j}{\partial x_i})\frac{\partial \tilde{u}_i}{\partial x_j}-\frac{2}{3}(\mu_t\frac{\partial \tilde{u}_k}{\partial x_k}+\bar{\rho}\tilde{k})\frac{\partial \tilde{u}_i}{\partial x_i}+C_{\varepsilon 3}\max(-g_i\frac{\mu_t}{\sigma_h}\frac{1}{\bar{\rho}}\frac{\partial \bar{\rho}}{\partial x_i},0))-C_{\varepsilon 2}\bar{\rho}\frac{\tilde{\varepsilon}^2}{\tilde{k}}+C_{\varepsilon 4}\bar{\rho}\tilde{\varepsilon}\frac{\partial \tilde{u}_i}{\partial x_i}$$

$$h=\sum_{n=1}^{N}Y_n h_n=\sum_{n=1}^{N}Y_n(h_{0,n}+\int_{T_0}^{T}c_{p,n}dT) \qquad \bar{\rho}=\frac{\bar{p}}{R\sum_{n=1}^{N}(\frac{\tilde{T}\tilde{Y}_n}{M_n})}$$

$$\bar{\tau}_{ij}=\mu(\frac{\partial \bar{u}_i}{\partial x_j}+\frac{\partial \bar{u}_j}{\partial x_i}-\frac{2}{3}\frac{\partial \bar{u}_k}{\partial x_k}\delta_{ij}) \qquad \bar{\rho}\widetilde{u_i^{"}u_j^{"}}=-\mu_t(\frac{\partial \tilde{u}_i}{\partial x_j}+\frac{\partial \tilde{u}_j}{\partial x_i})+\frac{2}{3}(\mu_t\frac{\partial \tilde{u}_k}{\partial x_k}+\bar{\rho}\tilde{k})\delta_{ij}$$

$$\bar{\rho}\widetilde{u_j^{"}h^{"}}=-\frac{\mu_t}{\sigma_h}\frac{\partial \tilde{h}}{\partial x_j} \qquad \bar{\rho}\widetilde{u_j^{"}Y_n^{"}}=-\frac{\mu_t}{\sigma_n}\frac{\partial \tilde{Y}_n}{\partial x_j}$$

$\sigma_k=1.0,\quad \sigma_\varepsilon=1.22,\quad \sigma_h=0.7,\quad \sigma_n=0.7,\quad C_{\varepsilon 1}=1.44,\quad C_{\varepsilon 2}=1.92,\quad C_{\varepsilon 3}=1.0,\quad C_{\varepsilon 4}=-0.33$

$n=\mathit{fuel},\mathit{oxygen},\mathit{product}$

Radiation plays an important role in the combustion process. It is solved here in parallel with the governing equations. The discrete transfer method (Lockwood et al, 1981) is used to provide the radiation source term for the energy equation of the gas phase and the radiation flux to the solid surface. The absorption and emission of gas and soot are also considered. The effects of gas and soot on the radiation intensity are given by the following equation:

$$\frac{dI}{ds}=-(k_g+k_s)I+\frac{\sigma}{\pi}(k_g+k_s)T^4$$

Scattering is negligible due to the small diameter of the soot particles. The absorption coefficients for gas and soot presented by Novozhilov et al. (1997) are used, and the soot density is assumed to be 2000kg/m^3 for the soot volume fraction (f_v) calculation (Jia et al, 1998).

$$k_g=0.28\exp(-\frac{T}{1135}) \qquad k_s=1264\,f_v T$$

The solution of these equations is based on the finite volume method. Convection terms are discretised using a second-order upwind differencing scheme. The SIMPLE pressure correction algorithm is used. The CFD model was validated with a fire wind tunnel experiment (Otake et al, 2004).

3.1.2. Firebrand Model

It is not reality to model all the physical processes of firebrand scattering at one time due to its complexity. Here we model its process following. Firebrands are assumed to be spherical, and the behavior of firebrands in the airflow can be considered to obey an air-solid two phase flow. Therefore, the scattering of firebrands can be presented by the Lagrangian transport equation. We assume that the combustion time of firebrands is much longer than the flying time, and the temperature of firebrands gives no influence on the air flow and the mass and shape are not changed during flight. From these assumptions we can obtain the following momentum equation:

$$m_d \frac{d\vec{u}_d}{dt} = \vec{F}_g + \vec{F}_{dr} + \vec{F}_p$$

where $\vec{F}_g = V_d(\rho_d - \rho_f)\vec{g}$ $(\vec{g} = (0,0,-9,8)m/s^2)$ is the force due to gravity. The drag $\vec{F}_{dr}$ is given by

$$\vec{F}_{dr} = \frac{1}{2} C_d \rho_f A_d |\vec{u} - \vec{u}_d| (\vec{u} - \vec{u}_d)$$

where $\vec{u}_d$ is the velocity of the firebrand, and $\vec{F}_p$ is the pressure gradient force due to the acceleration motion of the fluid (Hinze, 1975). This term is thought to be important because the fluid is accelerated by the thermal plume around the fire.

$$\vec{F}_p = -V_d \nabla p$$

The drag coefficient C_d is calculated using the following expression:

$$C_d = \begin{cases} 24(1+0.15\mathrm{Re}_d^{0.687})/\mathrm{Re}_d, & \mathrm{Re}_d \le 10^3 \\ 0.44 & , \mathrm{Re}_d > 10^3 \end{cases}$$

where Re_d is the Reynolds number of the firebrand.

$$Re_d = \frac{\rho_f |\vec{u} - \vec{u}_d| D_d}{\mu}$$

The air velocity $\vec{u}$ is given by the sum of the average velocity $\tilde{\vec{u}}$ and the turbulence fluctuation $\vec{u}'$.

$$\vec{u} = \tilde{\vec{u}} + \vec{u}'$$

The turbulence fluctuation is selected randomly from the Gaussian distribution probability density function with the following standard deviation σ.

$$\sigma = \sqrt{2k/3}$$

$$G(\vec{u}') = \frac{1}{\sqrt{2\pi}\sigma}\exp(-\frac{\vec{u}'^2}{2\sigma^2})$$

3.2. Measurement of Fluid Dynamics Properties of Firebrands

Firebrands have various complex shapes (spherical, cylindrical, and board shape, etc.), and it is very difficult to evaluate the drag force for all shapes. Then, firebrands are assumed to be spherical and the diameter which is used in the Lagrange equation is represented by the Stokes diameter of the firebrands here. Firebrands were collected from a fire wind tunnel experiment (Yoshioka, 2003) and the Stokes diameters of them are determined from their densities and the terminal velocities. The detailed measurement can be referred from Huang et al (2006). The comparison of the diameter (equivalent diameter) of the firebrands when it is assumed to be sphere to the Stokes diameter was done. The results showed that the ratio of Stokes diameter to equivalent diameter is focus on 0.2-0.7. This data was used for the simulation of firebrands scattering.

3.3. Real Urban Scale Fire Simulation

Figure 10 shows the calculation area which is X×Y×Z=510m×110m×200m. The size of each building is set to be 10m×10m×10m, and the building interval is also set to be 10m. There are fifteen buildings (5×3) and the building on fire is shown. The boundary conditions are shown in Table 4 and the simulation cases are shown in Table 5.

Table 4. Boundary conditions

Inlet	$U=U_D(z/10)^{1/4}$, T=293(K) $\varepsilon(z)=C_\mu^{3//4}k(z)^{3/2}/l(z)$ $l(z)=(4C_\mu k(z))^{1/2}D^{1/4}z^{3/4}/U_D$ k:Experimental data (Murakami,1988)
Outlet	Free flow
Wall	Generalized log law
Burning wall surface on fire	Inlet condition W_z=0.018m/s (Appendix A)

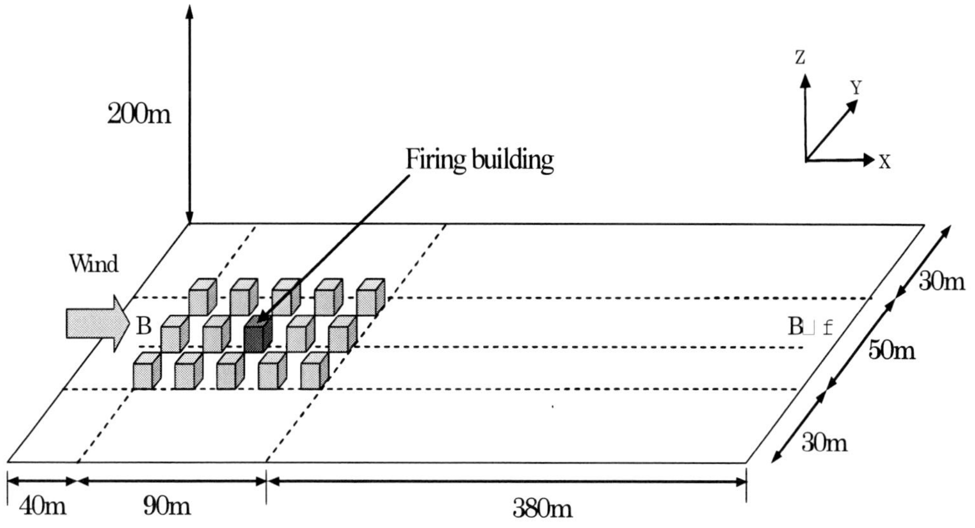

Figure 10. Analysis domain.

For all cases, the flow rate of the combustible gas (propane) is controlled from the burning face to get a heat release of 1.6MW/m^2 (Shiraishi et al., 2001).

Table 5. Simulation cases

	Inflow velocity (Height 10m)(m/s)	Burning wall on fire
CASE1	5m/s	roof
CASE2		Leeward side
CASE3		All sides
CASE4	10m/s	roof
CASE5		Leeward side
CASE6		All sides

*All sides include the roof.

The inflow wind velocity is assumed 5.0m/s and 10.0m/s based on the investigation of the Wakayama Shirahama Spa Fire that occurred in 1998 (Ohmiya et al., 1999), and the wind profile is given according to 1/4 profile for urban area. Because the firebrands about 1cm are most generated from past investigations, the cases that the equivalent diameter is 1.0cm and ratio of the Stokes diameter to the equivalent diameter is 0.2-0.7 were simulated. The firebrands are assumed to be generated from 81 points averaged over the burning face, and the interval between points is 1.0m. The firebrands that the ratio of the Stokes diameter to the equivalent diameter is 0.2, 0.3, 0.4, 0.5, 0.6, and 0.7 cm are generated at the same generation ratio from generation positions. There is almost no founding about the initial generation velocity of the firebrands. Here, we think that the generation velocity is close to the velocity of the fire plume in the top story of the building. It is reported that 5.0-10.0 Pa is the pressure difference between the fire room and the elevator shaft when the temperature is from 200°C to 800°C in building fire reported by Tamura (1994). We assume that the pressure difference between the fire plume and outdoor air is almost the same. Then, the velocity of the plume can be estimated to be 3.7-7.8m/s from the Bernoulli equation ($V = \sqrt{\frac{2\Delta p}{\rho}}$). Here, an average value, 5.0m/s is set to be the initial generation velocity here.

3.4. Results

3.4.1. Wind Distribution and Temperature Distribution

The vertical wind distribution and temperature distribution in the center section (section B-B') are shown in Figure 11 and Figure 12.

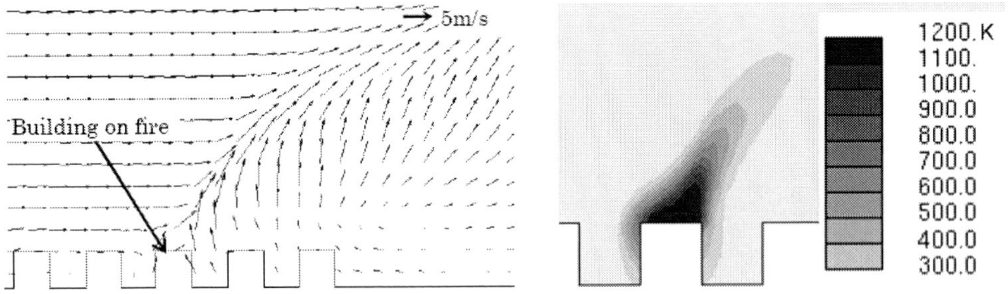

(1) CASE3 (Burning face: all sides)

Figure 11. Vertical wind and temperature distribution (Section B-B', Inflow velocity: 5m/s).

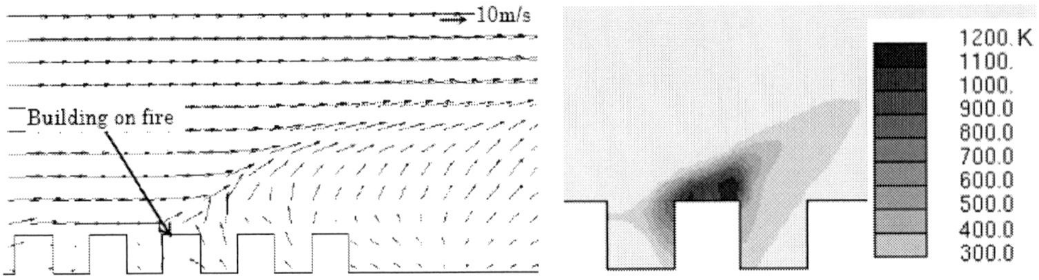

(1) CASE6 (Burning face: all sides)

Figure 12. Vertical wind and temperature distribution (Section B-B', Inflow velocity: 10m/s).

The thermal plume is strongest for the all burning cases, the one of roof fire is stronger than that of sidewall fire. The influence on the sky is little for the sidewall fire, though strong rising flow is seen between the buildings. When the inflow wind velocity is higher, the thermal plume is suppressed and greatly inclines to the leeward side, which increases the risk of the fire spread to the neighboring buildings.

3.4.2. Firebrands Scattering

Figure 13 shows the firebrands scattering results. The results show that firebrands can influence a wide area in each case. For the cases of roof fire, when the inflow wind velocity is 5.0m/s (CASE1), the firebrands with Stoke diameter of 0.2cm can scatter 400m or more, and the firebrands with Stoke diameter of 0.3cm can scatter about 100m. The firebrands with Stokes diameter of 0.4cm or more almost fall down at neighboring buildings. When the inflow velocity is 10.0m/s (CASE4), a part of the firebrands with Stoke diameter of 0.2cm scatter over 400m, and the firebrands with Stoke diameter of more than 0.3cm are all fall down at neighboring buildings. Wind velocity is strong in CASE4 compared with CASE1, so the rising flow is suppressed and scattering distances are shortened for all kinds of firebrands.

For the cases of sidewall fire, when the inflow velocity is 5.0m/s (CASE2), the firebrands with Stoke diameter of 0.2cm can scatter about 200m, and the firebrands with Stoke diameter of 0.3cm can scatter about 100m. the firebrands with Stokes diameter of 0.4cm or more almost fall down at neighboring buildings. When the inflow velocity is 10.0m/s (CASE5), the firebrands with Stoke diameter of 0.2cm can scatter about 150m, and the firebrands with Stoke diameter of 0.3cm can scatter about 80m..

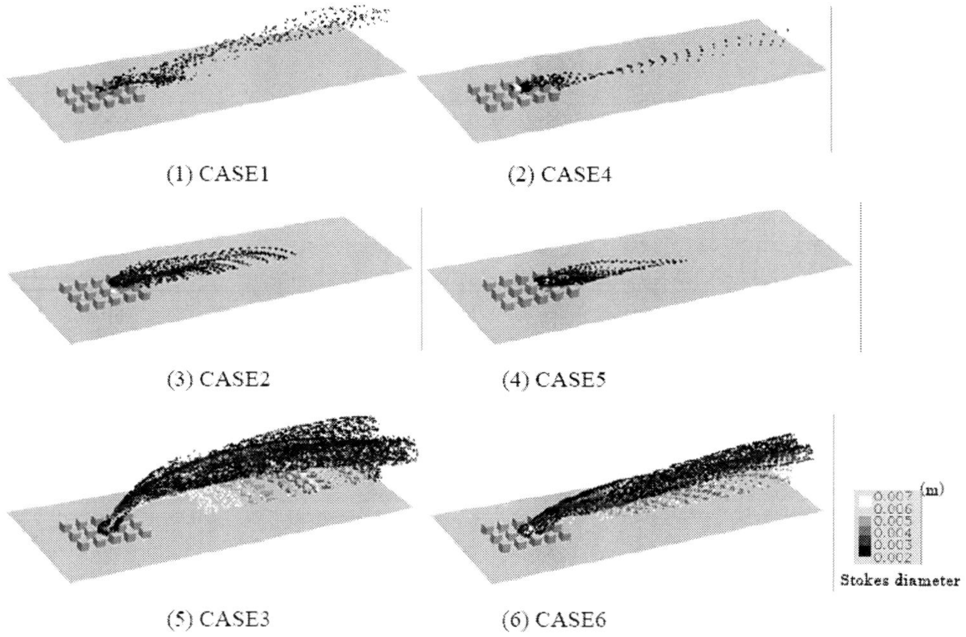

Figure 13. Firebrands trajectories.

The firebrands with Stokes diameter of 0.4cm or more almost fall down at neighboring buildings. The rising flow for CASE5 is weaker than CASE2 that the inflow wind velocity is low and scattering distances are shortened for all kinds of firebrands.

For the all sides burning cases, all kinds of firebrands scatter over 100m due to the rising flow. Comparing with CASE3 (inflow wind velocity: 5.0m/s), the scattering distances increase for CASE6 (inflow wind velocity 10.0m/s). This is because that though the rising flow in CASE6 is a little weaker, the firebrands can be raised up to the sky due to the rising flow, then follow the wind in the sky which is stronger in CASE6.

4. Conclusion

In this study, in order to study the urban thermal environment and urban safety, Computation Fluid Dynamics (CFD) modeling has been applied. At first, a numerical simulation program which can be adapted to complex urban areas coupled with convection, radiation and conduction is developed and used to predict the urban thermal environment. A comparison of the measured and simulated results for temperature, humidity, and wind velocity confirmed that this is an effective tool for assessing complex urban thermal environments. Moreover, it is thought that the cooling towers have little influence on the surrounding thermal environment at the pedestrian level because the sensible heat and the latent heat are expelled upwards into the sky. Secondly, CFD model, turbulent combustion model, radiation model and firebrand scattering model are coupled to predict and urban fire and the scattering of firebrands in a modeled urban area. It is found that when the inflow wind velocity is comparatively slow, the size of the thermal plume becomes significant, and when the inflow wind velocity is high, the thermal plume is suppressed and greatly inclines to the leeward side, which increases the risk of the fire spread to neighboring buildings. The firebrand can scatter over a wide area more than 400m. Firebrands are blown up by the thermal plume and the scattering distances are influenced by the background flow field, the magnitude of thermal plume, and the Stokes diameters of firebrands.

Appendix A

The boundary condition of burning wall surface which is on fire is set to be inlet condition. W_z means the inflow velocity of propane perpendicular to the wall surface. It is decided as follow:

We know that $Q = \rho_{C3H8} W_z A \Delta H_C$

Q: Heat release rate (kW).
ρ_{C3H8}: Density of propane (1.9 kg/m^3)

A: Area of burning wall surface (m^2)

ΔH_C: Heat of Combustion of propane (46,460 kJ/kg)

Then, we can calculate W_z from the heat release rate.

References

Chen, H., Ooka, R., Harayama, K., Kato, S. Li, X. 2004. Study on outdoor thermal environment of apartment block in Shenzhen, China with coupled simulation of convection, radiation and conduction, *Energy and Buildings*, **36**, 1247-1258.

El Tahry, S. H. 1983. k-ε Equation for compressible reciprocating engine flows, *AIAA J. Energy*, 7(4): 345-353.

Harayama, K., Yoshida, S., Ooka, R., Mochida, A., Murakami, S. 2002. Prediction of outdoor environment with unsteady coupled simulation of convection, radiation and conduction, Part 1, Numerical study based on unsteady radiation and conduction analysis, *Journal of Architecture, Planning and Environmental Engineering* **556**, 99-106. (in Japanese).

Hinze, J. O. 1975. *Turbulence*, McGraw-Hill, New York, 460-471.

Huang, H., Ooka, R., Kato, S. 2005. Urban thermal environment measurements and numerical simulation for an actual complex urban area covering a large district heating and cooling system in summer, *Atmospheric Environment*, **39**, 6362-6375.

Huang, H., Ooka, R., Kato, S., Hayashi, Y. 2006. A numerical study of firebrands scattering in urban fire based on CFD and firebrands aerodynamics measurements, *Journal of Fire Sciences*, accepted.

Iwami, T. and Koji, K. 2002. Report on city fire in Wakkanai-Part 2, State of damaged buildings and distribution of firebrands in the surrounding area, *Fire*, **52**: 51-53. (In Japanese)

Jia, F., Galea, E. R. and Patel, M. K. 1998. The numerical simulation of fire spread within a compartment using an integrated gas and solid phase combustion model, *Journal of Applied Fire Science*; **8**(4):327-352.

Lockwood, F. C. and Shah, N. G. A. 1981. New radiation solution method for incorporation in general combustion prediction procedures, *Proceeding of 18th Symposium (International) on Combustion,* The Combustion Institute, 1405-1414.

Magnussen, B. F., Hjertager, H. 1976. On mathematical modeling of turbulent combustion with special emphasis on soot formation and combustion. *Proceeding of 16th Symposium (International) on combustion*, The Combustion Institute, 719-729.

Murakami, S., Mochida, A., Hayashi, Y. 1988. Modification of production terms in k-ε model to remove overestimate of k value around windward corner, *10th Wind Engineering Symposium 199-204.* (in Japanese).

Novozhilov, V., Harvie, D. J. E., Kent, J. H., Apte, V. B. and Pearson, D. A. 1997. Computational fluid dynamics study of wood fire extinguishment by water sprinkler, *Fire Safety Journal,* **29**:259-282.

Novozhilov, V., Moghtaderi, B., Fletcher, D. F. and Kent, J. H. 1996. Computational fluid dynamics modeling of wood combustion, *Fire Safety Jounal*, **27**:69-84.

Ohmiya, Y. and Iwami, T. 1999. An investigation on the distribution of firebrands and spot fires in a hotel fire, *AIJ Journal of Technology and Design*, **9**: 133-136. (In Japanese).

Otake, H., Huang, H., Ooka, R., Kato, S., Hayashi, Y. 2004. Simulation of flames and thermal plume in urban fire under windy condition, *Journal of Institute of Industrial Science*, **56**:11-16. (In Japanese).

STAR-CD. 2001. *Methodology*, STAR-CD Version 3.15. Computational Dynamics Limited.

Shiraishi, Y., Kato, S., Yoshida, S. and Murakami, S. 2001. Numerical analysis of scatter of firebrands within urban fire spread, *Journal of Architecture, Planning and Environmental Engineering*, **546**: 187-192. (In Japanese).

Tamura, G. T. 1994. *Smoke movement and control in high-rise buildings*, The National Fire Protection Association, Quincy, Massachusetts.

Yan, Z. H. and Holmstedt, G. 1996. CFD and experimental studies of room fire growth on wall lining materials, *Fire Safety Journal*, **27**:201-238.

Yoshida, S., Murakami, S., Mochida, A., Ooka, R., Tominaga, Y., Kim, S. 2000a. Influence of green area ratio on outdoor thermal environment with coupled simulation of convection, radiation and moisture transport, *Journal of Architecture, Planning and Environmental Engineering* **529**, 77-84. (in Japanese).

Yoshida, S., Ooka, R., Mochida, A., Tominaga, Y., Murakami, S. 2000. Study on effect of greening on outdoor thermal environment using three dimensional plant canopy model,

Journal of Architecture, Planning and Environmental Engineering **536**, 87-94. (in Japanese)

Yoshioka, H., Hayashi, Y., Omiya, Y., Noguchi, T., Kato, S. and Ooka, R. 2003. Full-scale fire wind tunnel experiments on generation of firebrands from a burning house, *Summaries of Technical Papers of Annual Meeting of Architectural Institute of Japan*, 117-118. (In Japanese).

In: Focus on Urbanization Trends
Editor: Laura M. De Smet
ISBN: 978-1-60456-109-8

URBANIZATION: 21ST CENTURY HEALTH ISSUES AND CHALLENGES IN THE NORTH WEST PROVINCE, SOUTH AFRICA

J.M. van Rooyen
School of Physiology, Nutrition and Consumer Sciences,
North-West University, Potchefstroom Campus,
Potchefstroom, South Africa

Abstract

People in South Africa especially the black population (Africans) are subjected to massive urbanization since the early 1990's. They migrate from the traditional rural settings where they live under the auspices of a tribe and tribal habits to the large peri-urban informal settlements near all the larger cities and towns in South Africa in order to seek other opportunities to supply in their daily subsistence.

This has an impact on their health and add to the chronic diseases of lifestyle risk factors typical of the western lifestyle they have to adopt. They are eating more salt, saturated fats, and refined sugars, although high in energy compared to the traditional rural diet. Their intake of fiber and potassium are lower because they are eating less fruit and vegetables than their rural counterparts. They are more physical inactive than their rural counterparts, they walk less because they are making use of public transport and spending more time watching television. Migration to an urban settlement is a stressful life event and this daily experienced stress of poor living conditions leading to more violence, drug and alcohol abuse which may contribute to the larger incidence of cardiovascular diseases amongst them. The poor diet most pregnant women follows are leading to stunted offspring. All this contribute to the higher risk for chronic diseases and diseases of lifestyle like hypertension, diabetes, and stroke. Stunting in children may also lead to hypertension and obesity in later life. The prevalence of most of this mentioned diseases of lifestyle are much higher in the urbanized subjects compared to their rural counterparts.

The major health burdens in South Africa are: infectious diseases like HIV/AIDS and tuberculosis, diseases related to malnutrition and chronic diseases as a cause of urbanization like hypertension and stroke a forerunner of vascular disease. The challenges would be a complementary dual approach to promote a healthy lifestyle for the whole population and the early diagnosis of disease in individuals with the cost-effective management of the risk factors and disease.

Introduction

People in South Africa especially the Africans are subjected to massive urbanization since the early 1990's. They migrate from the traditional rural settings where they live under the auspices of a tribe and tribal captain to the peri-urban informal settlements near all the larger cities and towns in South Africa in order to seek other opportunities to supply in their daily subsistence. Since 1970 to 1994, the world rates of urbanization increased from 36.6% to 44.8% and it is expected to increase further to 61.1% in the year 2025 [1]. The same trends are found in other developing countries and increased from an urbanization rate of 12.6% in 1970 to 21.9% in 1994 [1]. It is not only the migration of individuals from rural to urban areas that take place, but the rural areas themselves are being transformed [1], for example in the rural areas of the North West province of South Africa the Coca-Cola sign is seen everywhere, they sell liquor from a liquor shop and cell phone reception towers are seen in the rural areas. Most households have a television and this will influence their subsistence. However there are still differences in lifestyle that exists between urbanized and rural people, like the rural people are still under the auspices of a tribal captain and they have support from their family members and friends. The urbanized lifestyle has an impact on the health of the urbanized people and add to the chronic diseases of lifestyle risk factors typical of the western lifestyle they have to adopt.

Because the Africans are the most vulnerable and subjected to massive transition and urbanization, this chapter will deal with the health effects as found in the African people of the North West province of South Africa as a cause of urbanization and the challenges for the 21st century.

Issues

Diet

A typical male rural African and a typical urban township diet are shown in table 1.

Table 1. The distribution of macronutrients in the diet of black South African males residing in different locations [2]

Dietary factor	WHO Goals [3]	Rural n=194	Farm workers n=108	Urbanized Informal Settlement n=128	Middle class n=229	Upper class n=83
Energy (MJ)		9.6	8.9	9.3	9.9	8.3
Total fat (% of energy)	15-30	22.9	22.8	24.3	26.0	30.6
SFA (%)	<10	16.1	16.9	16.8	19.0	24.9
PUFA's (%)	6-10	14.8	12.2	14.6	16.5	17.2
CHO (%)	55-75	67.4	67.2	65.5	64.0	57.3
Protein (%)	10-15	11.6	12.1	12.0	11.8	13.2
Cholesterol (mg)	≤300	315	283	332	377	420
Fiber (g)		19.2	15.6	17.4	18.8	19.7

SFA = fatty acids, PUFA's = poly-unsaturated fatty acids, CHO = carbo hydrates

The rural diet consist of a low fat and high carbohydrate intake, which is also typical of a diet found in other African countries [2,4]. They also have a low intake of fruit and vegetables (134.2 g versus 400 g recommended by the WHO [3,5] especially in the higher class of urbanized people [2]. The mean energy distribution was 65% carbohydrate, 12% protein and 22% fat for the rural versus 57% carbohydrate, 13% protein and 31% fat for the urbanized subjects [2].

Physical Inactivity

In a study done in the North West province of South Africa in a rural and an urbanized community [6] more than half of the subjects participated in the study were not active enough [6]. In table 2 the prevalence of physical inactivity in an adult sample of South Africans is shown. In another study conducted between 2002 and 2003 as a representative sample (n=2014) of urban and rural South Africans [7] shows that less than one third of South Africans met the American College of Sports Medicine and Centers for Disease Control recommendation for health-enhancing physical activity (subjects must accumulate 30 minutes of moderate activity on most, but preferably all days of the week), and 46% were inactive [7].

Table 2. Prevalence of physical inactivity in a representative sample of adult South Africans [7], 95% Confidence Intervals shown in brackets

	Men	Women	All
Inactive (<600 MET min/wk)	43 (38;49)	49 (43;54)	46 (42;51)
Minimally Active (≥600 MET min/wk)	20 (16;23)	27 (23:30)	24 (21;27)
Sufficiently Active (HEPA)	37 (32;42)	25 (20;29)	30 (26;34)

HEPA (Health enhancing physical activity; ≥ 7 days of any combination of moderate and vigorous activity, ≥ 3000 MET min/wk).

Twenty nine percent of the subjects were classified as inactive and 28% as moderately active [6,7]. The men were significantly more active than the women. Inactive overweight men and women had the highest mean systolic blood pressure, total serum cholesterol and low-density lipoprotein cholesterol [6].

Daily Stress

Urbanization may lead to social and cultural disruption causing increased levels of stress [8,9,10]. During daily stress situations like subsistence, the urbanized people are subjected to, are also causing greater cardiovascular reactivity which may lead to the development of hypertension [9, 11] in later life.

Children living in informal housing, in urbanized areas near the larger cities in South Africa are subjected to violence and the violence exposure manifests in psychological disorders such as anxiety, depression, fear, helplessness and hopelessness, emotional withdrawal and somatic symptoms [12]. The higher levels of environmental stress the black children are exposed to [12] could change their vascular sensitivity to sympathetic stimulation

[12,13]. This may increase the peripheral vascular resistance directly and contribute to the pathogenesis of hypertension in their later lives.

Further statistically significant correlations were found in adults (men and women) between some psychological variables and blood pressure [8], see Table 3.

Table 3. Spearman correlations between psychosocial variables and blood pressure (adjusted for all variables which are independently associated with blood pressure[x]), n=814

Psychosocial variables	Men				Women			
	SBP		DBP		SBP		DBP	
	R	P	R	P	R	P	R	P
SACS:								
ASA		NS		NS	.13	.016	.12	.031
AGA		NS		NS		NS		NS
SJ		NS		NS	0.12	.03	.18	.001
SSS		NS		NS		NS	.12	.03
PSS:								
FR		NS		NS	-.11	.03	-.13	.02
FA		NS		NS	-.09	.05	-.13	.02
Tot		NS		NS		NS	-.13	.02
AAS:								
RBP	.10	.05	.11	.04		NS		NS
PAT		NS		NS		NS		NS
Psychosocial variables	**Men**				**Women**			
	SBP		**DBP**		**SBP**		**DBP**	
	R	**P**	**R**	**P**	**R**	**P**	**R**	**P**
I - C:								
FI		NS	.10	.05				
SWH		NS		NS		NS		NS
AFM:								
NA	.14	.01		NS		NS		NS
PA		NS		NS		NS		NS

SACS = Strategic Approach to Coping Scale, ASA = Anti Social Action, AGA = Aggressive Action, SJ = Social Joining, SSS = Seeking Social Support; PSS = Perceived Social Support, FR = Friends, FA = Family, Tot = Total Score; I - C = Individualism - Collectivism Scale; AAS = African Acculturation Scale; RBP = Religious Beliefs and Practices, PAT = Preference for African Things; FI = Family Integrity, SWH = Self Reliance with Hedonism; AFM = Affectometer, NA = Negative Affect, PA = Positive Affect [8]. [X] Age and stratum of urbanization in men, age, stratum, BMI, WC, snuff taking, TC, LDL-cholesterol and log fasting insulin in women.

The psychological variables associated with blood pressure in the subjects subjected to transition were: coping strategies (SACS), perceived social support from family and friends (PSS), aspects of acculturation (AAS), and individualism vs. collectivism (I-C), as well as prevalence of negative or positive affect (AFM) [8]. The pattern of correlations differ in men and women [8]. Apparently other dimensions of psychosocial stress which were not measured could also be involved [8]. It should be mentioned that the social support in urbanized subjects are collapsing and could lead to vicious circle of more stress.

Blood Pressure and Pulse Wave Velocity (PWV)

The study sample shown in Table 4 consisted mainly of Setswana speaking subjects (73.3%) and the rest of the sample consisted of Sotho (11%), Xhosa (10%), Zulu and Shangaan (5.7%) subjects [8] living in the North West province of South Africa.

The influence of tribal heterogeneity on blood pressure is unknown, however, 22.8% of the subjects of the total sample had systolic, and 20.7% diastolic blood pressures above 140/90 mmHg on the day of the study. Males and females from stratum 1 to 4 (that is rural to urbanized) show increases in the incidence of high blood pressure [8]. The rate of mild hypertension in stratum 3 was 32.9% for systolic and 25.1% for diastolic pressure respectively [8]. In the African population of the North West province of South Africa the higher blood pressure found in urbanized compared to rural subjects could be ascribed to endothelial dysfunction and not to atherosclerosis because the BMI's and lipid profiles were the same in both rural and urbanized subjects [14]. The PWV (an index of stiffness in the elastic arteries) in African male subjects is not influenced by urbanization when the PWV of rural and urbanized males were compared [15]. This explains the results that the higher blood pressure as found in the urbanized Africans is rather ascribed to endothelial dysfunction than atherosclerosis or changes in stiffness of the blood vessels.

Table 4. The rate of hypertension expressed as a percentage of subjects (n=1783) who have SBP and DBP greater than 140/90 mmHg in the different strata [8] of urbanization (stratum 1, deep rural to stratum 3,4 and 5, urbanized)

Gender	Variable	Stratum*1	Stratum 2	Stratum 3	Stratum 4	Stratum 5	Total
Male	n	198	114	132	232	84	760
	SBP	19.2	13.2	34.8	23.3	17.9	22.1
	DBP	14.6	13.2	22.7	19.0	22.6	18.0
Female	n	299	149	175	294	106	1023
	SBP	22.1	28.2	31.4	31.3	10.4	26.0
	DBP	21.4	21.5	26.9	25.9	13.2	22.8
Total	SBP	20.9	21.7	32.9	27.8	13.7	22.8
	DBP	18.7	17.9	25.1	22.8	17.4	20.7

* Strata: stratum 1 = deep rural, stratum 2 = farm-workers, stratum 3 = living in informal settlements, stratum 4 = living in established middle-class townships, stratum 5 = living in upper-class urban areas [8].

Stunting in Children

Stunting during childhood increases the risk of non-communicable diseases in adulthood as reviewed by Vorster [16]. It is therefore feasible to study nutritional disorders like stunting in children in order to establish the role in future health events. In a study sample of children, evenly distributed within the age limits of 10-15 years and stratified for age and living conditions, 35.8% of them lived in rural communities and 46.4% lived in urban areas [17]. The prevalence of stunting was 26% amongst the African boys and 21% of African girls were stunted aged 10-15 years [17]. At the time of this study no differences existed between dietary intakes of stunted and non-stunted children because of fortification of staple foods and feeding schemes in schools [17]. Evidence supports Barker's hypothesis that it is poor or imbalanced delivery of nutrients in utero to the fetus that programs high blood pressure in humans [17]. A vicious circle of events is proposed that leads to a lower arterial compliance in stunted children which originated in parental under nutrition in areas of low socioeconomic support, which leads to children with low birth weight, undernourished and stunted who will develop hypertension in later life [17].

Endocrine Parameters Associated with Urbanization

Lower cortisol and testosterone levels and higher prolactin levels were found in the urbanized compared to the rural African females. The prolactin levels were higher in the urbanized males but the cortisol and testosterone levels did not differ in urbanized and rural African males. The urbanized subjects show an endocrine pattern of ongoing stress that can be associated with urbanization [18]. Increased levels of cortisol are expected in the traditional view when the HPA axis is stimulated, but recently hypocortisolism is reported under conditions of chronic stress probably as a result of increased feedback inhibition of the pituitary-adrenal level of the HPA axis [19]. This explains the lower cortisol found in the African urbanized females because they are more under stress than the males to supply food and participate in generating income. Heightened prolactin levels as well as low levels of testosterone may disturb the delicate balance between vasoconstriction and vasodilatation in favor of constriction [18] and this may contribute to higher blood pressure levels as found in the urbanized subjects [8].

Challenges for the 21st Century

The major health burdens in South Africa are: infectious diseases like HIV/AIDS and tuberculosis, diseases related to malnutrition and chronic diseases as a cause of urbanization like hypertension and stroke a forerunner of vascular disease. The challenges would be a complementary dual approach to promote a healthy lifestyle for the whole population and the early diagnosis of disease in individuals with the cost-effective management of the risk factors and disease.

Challenges Regarding Child Health

The major health problems in children in South Africa are: under-nutrition, preventable chronic childhood infections such as diarrhea and respiratory infections, emerging epidemics including HIV/AIDS and the scourge of violence and trauma against children [20]. The United Nations at their June 2000 summit meeting had identified a number of global priorities for children, including child nutrition, HIV/AIDS, trauma, violence against children and social security for children [20]. The South African priorities for urbanized children are similar.

Challenges Regarding Prevention Strategies

The promotion of healthy food especially for the poorer urbanized communities near the larger cities must be a priority [5]. The salt content in staple foods like bread and processed food in South Africa is high, low salt alternatives should be developed [5]. Because older persons, African Americans, and possibly women are more responsive to sodium/salt reduction [21], it should also be advantageous to the South African population to reduce sodium intake. High levels of physical inactivity is a concern and children should be encouraged to do more exercise. The sport administrators involved with soccer should take the advantage of world cup soccer scheduled for 2010 in South Africa to encourage school children to participate more in organized sport. This will enhance their health and will keep them away from practicing crime and violence as it is evident from the informal settlements (urbanized communities) near the larger cities. There is a very high prevalence of obesity among black South African women [22, 23, 24] and it seems that many black women do not really want to loose weight [22] because obesity is culturally and aesthetically looked upon [22]. Holistically seen, indicators revealed that urban obese women showed the most adverse health profile compared to lean women [24]. This accounts for their exposures, namely low socio-economic status, health risk behaviors such as alcohol intake, and their health-related outcomes which include cardiovascular and endocrine data, triglyceride concentrations, haemostatic conditions and their level of psychological well-being [24]. Public-health programs for the prevention and treatment of obesity and weight loss among black women should include a component of increased physical activity such as culture-sensitive dancing exercises [22]. Culturally congruent health information should be made available and the level of education of the target audience should be taken in consideration [5]. People should be make more aware about their own and family medical history, including the known risk factors for cardiovascular diseases like hypertension, diabetes, overweight [5] and stroke.

Challenges Regarding Prevention and Treatment of Hypertension

The promotion of health should not only target the broad population but should also target the individual. The overall management of hypertension is as much a socioeconomic as it is a therapeutic problem [25]. Screening individuals will be ideally not only to detect hypertension but is also the basis for education and therapy [25]. The detection and management of hypertension are such a vast problem and becoming a political issue because of limited financial resources [25]. Policy makers and government (local, regional and national) should make wider policies in this regard.

Challenges Regarding Social Support

Psychological parameters like coping strategies and perceived social support from family and friends correlated with blood pressure [8] it must be a challenge for governments and policy makers in the 21st century to promote social support because to our experience urbanized subjects lacking social support and have high blood pressures [8]. Individuals with high cultural consonance (the degree to which individuals are able to approximate in their own behaviors the prototypes for behavior encoded in shared cultural models) in the domains of lifestyle and social support had lower blood pressures [26]. Cultural dimensions of everyday life for health are independent of other social, psychological and dietary factors [26]. Social workers should undertake projects in urbanized communities to make them aware of the lack in social support and encourage them to socialize.

Challenges Regarding Ingestion of Indigenous Food

When people migrate from the traditional rural lifestyle to the urbanized lifestyle their traditional eating habits change and knowledge regarding traditional food disappear. Therefore for example the decline in consumption of *morogo* (*morogo*= Setswana word, which is the collective term for dark green leafy traditional vegetables utilized by African communities) vegetables and eventually exclusion of these rich sources of ω-3 Poly-unsaturated fatty acids (PUFAs) and folate from the diet could be an important factor contributing to the high prevalence of lifestyle diseases in urbanized societies of African origin [27]. Regular *morogo* consumption would increase intake of ω-3 PUFAs and folate because these plants are rich in fatty acids and folate. A considerable variation in folic acid concentrations in *morogo* was found which varies from 72 μg / 100g fresh weight (amaranth, Rustenburg, South Africa) to as high as 217 μg / 100g (African cabbage, Capricorn) [27]. A regular consumption of *morogo* may contribute towards a more appropriately balanced ω-6:ω-3 PUFA ratio that would strengthen genetically-predisposed mechanisms protecting black South Africans [27] from the development of cardiovascular diseases and overall consumer health [28]. It should be a challenge to encourage urbanized African people and make them aware of the advantage of eating their traditional African food. Up to know rural Africans collect some plants that grow wild, however small farmers could be encouraged to grow these plants from their own region to supply the supermarkets.

References

[1] Yusuf, S; Reddy, S; Ôunpuu, S; Anand, S. Global burden of cardiovascular diseases: Part I: General considerations, the epidemiologic transition, risk factors, and impact of urbanization. *Circulation*, 2001, 104, 2746-2753.

[2] Maclntyre, UE; Kruger, HS; Venter, CS; Vorster, HH. Dietary intakes of an African population in different stages of transition in the North West Province, South Africa: the THUSA study. *Nutr. Res.*, 2002, 22, 239-256.

[3] WHO/FAO Expert Consultation. Diet, Nutrition, and the Prevention of Chronic Diseases. Technical Report Series 916. Geneva: World Health Organization, 2003.

[4] Tanner, M; Lukmanji, Z. Food consumption patterns in a rural Tanzanian community. *Acta Trop.*, 1987, 44, 229-244.

[5] Chronic Diseases of Lifestyle in South Africa: 1995-2005. *Medical Research Council – Technical Report.* (Eds.) Steyn, K; Fourie, J; Temple, N. May, 2006.

[6] Kruger, HS; Venter, CS; Vorster, HH. Physical inactivity as a risk factor for cardiovascular disease in communities undergoing rural to urban transition: the THUSA study. *Cardiovasc. J. S. Afr.,* 2003, 14, 16-23.

[7] World Health Organization. Preliminary results of the World Health Survey, 2002-2003; *International Physical Activity Data*, South African results. Geneva: WHO, 2005.

[8] Van Rooyen, JM; Kruger, HS; Huisman, HW; Wissing, MP; Margetts, BM; Venter, CS; Vorster, HH. An epidemiological study of hypertension and its determinants in a population in transition: the THUSA study. *J. Hum. Hypertens.*, 2000, 14, 779-787.

[9] Malan, NT; Van der Merwe, JS; Huisman, HW; et al. A comparison of cardiovascular reactivity of rural blacks, urban blacks and whites. *Stress Med.*, 1992, 8, 241-246.

[10] Aldrich, RA. The social context of change. *Psych. Ann.*, 1986, 16, 613-618.

[11] Van Rooyen, JM; Huisman, HW; Eloff, FC; Laubscher, PJ; Malan, L; Malan, NT. Cardiovascular reactivity in black South-African males of different age groups: The influence of urbanization. *Ethn. and Dis.*, 2002, 12(1), 69-75.

[12] Van Rooyen, JM; Nienaber, AW; Huisman, HW; Schutte, AE; Malan, NT; Schutte, R; Malan, L. Differences in resting cardiovascular parameters in 10- to 15-year old children of different ethnicity: The contribution of physiological and psychological factors. *Ann. Behav. Med.*, 2004, 28(3), 163-170.

[13] Calhoun, DA. Hypertension in blacks: Socioeconomic stress and sympathetic nervous system activity. *The Am. J. of the Med. Sci.,* 1992, 304(5), 306-311.

[14] Schutte, R; Huisman, HW; Malan, L; Van Rooyen, JM; Schutte, AE; Malan, NT; De Ridder, JH. Differences in cardiovascular function of rural and urban African males: the THUSA study. *Cardiovasc. J. SAfr.*, 2004, 15(4), 161-165.

[15] Van Rooyen, JM; Huisman, HW; Schutte, AE; Fourie, CMT; Malan, NT; Malan, L; Schutte, R; Palmer, L. A comparison of pulse wave velocity and blood pressure in men and women from African descent. *J. Hypertens.,* 2007, 25(suppl.2), S347.

[16] Vorster, HH. The emergence of cardiovascular disease during urbanization of Africans. *Public Health Nutr.*, 2002, 5, 239-243.

[17] Van Rooyen, JM; Kruger, HS; Huisman, HW; Schutte, AE; Malan, NT; Schutte, R. Early cardiovascular changes in 10-15-year-old stunted children: the Transition and Health during Urbanization in South Africa in Children study. *Nutrition*, 2005, 21, 808-814.

[18] Huisman, HW; van Rooyen, JM; Malan, NT; Eloff, FC; Malan, L; Laubscher, PJ; Schutte, AE. Prolactin, testosterone and cortisol as possible markers of changes in cardiovascular function associated with urbanization. *J. Hum. Hypertens.,* 2002, 16, 829-835.

[19] Heim, C; Ehlert, UM; Hellhammer, DH. The potential role of hypocortisolism in the pathophysiology of stress-related bodily disorders. *Pshychoneuroendocrin.*, 2000, 25, 1-35.

[20] Shung-King, M; Abrahams, E; Giese, S; Guthrie, T; Hendricks, M; Hussey, G; et al. *Child Health- South African Health Review 2000*. Durban, South Africa: Health Systems Trust.

[21] Whelton, PK; He, J; Louis, GT., (Ed.), *Lifestyle Modification for the Prevention and Treatment of Hypertension*, Marcel Dekker, Inc.,New York, 2003.

[22] Kruger, HS; Venter, CS; Vorster, HH; Margetts, BM. Physical inactivity is the major determinant of obesity in black women in the North West province, South Africa: the THUSA study. *Nutrition*, 2002, 18, 422-427.

[23] Steyn, K; Jooste, PL; Bourne, LT; et al. Risk factors for coronary heart disease in the black population of the Cape Peninsula. The BRISK study. *S. Afr. Med. J.*, 1991, 79, 480.

[24] Schutte, AE; Kruger, HS; Wissing, MP; Underhay, C; Vorster, HH. The emergence of the metabolic syndrome in urban obese African women: the POWIRS study. *S. Afr. J. of Sci.*, 2005, 101, 61-67.

[25] Opie, LH; Seedat, YK. Hypertension in Sub-Saharan African populations. *Circulation*, 2005, 112, 3562-3568.

[26] Dressler, WW; Balieiro, MC; Ribeiro, RP; Dos Santos, JE. Cultural consonance and arterial blood pressure in urban Brazil. *Social Sci. and Med.*, 2005, 61, 527-540.

[27] Van der Walt, AM; Loots, Du T; Ibrahim, MIM; Bezuidenhout, CC. Alpha-linolenic acid and folate in traditional African leafy vegetables and possible health benefits to an urban black South African population. (Unpublished results 2007).

[28] Van der Walt, AM; Mossanda, KSA; Jivan, SD; Swart, WJ; Bezuidenhout, CC. Indigenous African food plants: vehicles of disease or sources of protection. *Indilinga, Afr. J. of Indigenous Knowledge System*, 2005, 4(1), 270-297.

RESEARCH AND REVIEW STUDIES

In: Focus on Urbanization Trends
Editor: Laura M. De Smet
ISBN: 978-1-60456-109-8

Chapter 1

URBANIZATION AND THE INCREASING RISK FROM MOSQUITO-BORNE DISEASES: LINKING HUMAN WELL-BEING WITH ECOSYSTEM HEALTH

***Paul T. Leisnham*[1*] *and David P. Slaney*[2]**
[1]Behavior, Ecology, Evolution, and Systematics Section
Department of Biological Sciences
Illinois State University
Campus Box 4120, Normal, IL, 61790, USA
[2]Institute of Environmental Science and Research Ltd
PO Box 50-384, Porirua, New Zealand

Abstract

Urbanization is considered one of the primary drivers of ecological change because it alters the basic physical properties of the environment, including the vegetation, hydrology, soil structure, and topography. These changes in turn alter the complex abiotic (physical and chemical) and biotic processes that typified the original ecosystem. This often leads to the elimination or severe disruption of the ecological niches of resident species, whilst new ecological niches may be formed. The world is experiencing resurgence in infectious diseases attributable to urbanization and associated environmental changes, marking a third epidemiological transition. The most pre-eminent vector borne diseases to re-emerge and coincide with urbanization have been those vectored by mosquitoes. Urbanization has frequently increased mosquito-borne disease risk directly by providing dense numbers of susceptible human hosts, encroaching on areas of high mosquito density, and creating favorable adult and larval habitats. Urbanization has also increased mosquito-borne disease risk indirectly by causing changes (usually reductions) to the resident biodiversity that facilitate the increase and spread of native and invasive species. Today, New Zealand provides a case study nation where urbanization may facilitate an economic and public health catastrophe from mosquito-borne human disease. Despite no historical precedent, recent studies suggest that urbanization is increasing the risk of mosquito-borne human disease in

[*] E-mail address: pleisnh@ilstu.edu. Phone: 309 438 2603. Fax: 309 438 3722. (Corresponding author: Dr Paul Leisnham)

New Zealand. Urbanization appears to be fostering the spread of resident mosquito species and increasing the likelihood of invasion from exotic vectors through its synergism with other anthropogenic changes, including climate change and increased international travel and trade, as well as natural factors, such as a depauperate native invertebrate fauna. Studying invasion processes help us understand how ecological shifts accompanying urbanization can increase the risk from disease vectors. This in turn will improve the surveillance and control of current exotic species, identify future successful invaders, and, ultimately, help limit the spread of mosquito-borne diseases. The proliferation of mosquito-borne diseases by urbanization highlights the intrinsic two-way link between ecosystem and human health. While traditional disease models can cope with the direct effect of an urban-mediated exposure for a specific disease outcome, the urbanization-mosquito relationship demands a multidisciplinary approach to understand the complex indirect ecological mechanisms leading to human disease.

Introduction

In presenting the Millennium Report in 2000, Secretary-General of the United Nations, Kofi Annan, ushered in the new age by reflecting upon pressing challenges facing humanity (Annan 2000). One of the most pre-eminent challenges outlined in the Report is the increasing burden degraded ecosystems are placing on human survival and well-being. The Millennium Report recognized the broader acceptance among health and environment experts and governments that there is an intrinsic two-way link between ecosystem resources and human health (McMichael and Haines 1997; VanLeeuwen et al.,, 1999; Haines et al.,, 2000; Chivian 2001; Millennium Ecosystem Assessment 2005; Weinstein 2005; Pongsiri and Roman 2007). Human activities can profoundly affect the ecosystem, while humans in turn depend on proper ecosystem functioning to provide the necessary services to maintain good health (Forget and Lebel 2001; Parkes and Weinstein 2004).

One of the primary drivers of decreased ecosystem functioning facing the 21st century is accelerating urbanization. The 2005 Revision of the UN World Urbanization Prospects Report stated that the 20th century observed "the rapid urbanization of the world's population", as the percentage of people in urban areas rose dramatically from 13% (220 million) in 1900, to 29% (732 million) in 1950, to 49% (3.2 billion) in 2005 (United Nations Population Division 2006). The same report projected that the figure is likely to rise to 60% (4.9 billion) by 2030. Urbanization can be defined as the increase over time in the population of cities in relation to the region's rural population. Greater human population density results in substantial landscape changes characterized by industry, services, transportation, built-up structures, goods, and waste. There are two main forms of urbanization that can increase the population of cities. The first form involves an increase in population density within a unit area, thus concentrating people and their associated activities. The second form involves the expansion of existing urban landscapes, which leads to an overall increase in human population but a relatively smaller increase in population density. Often, the growth of cities involves both forms of urbanization acting simultaneously. Usually, however, the relative contributions of these forms are not evenly distributed in the global context. Increasing population density is the predominant form of urbanization in developing nations, since the majority of the population still live in rural areas (United Nations Population Division 2006). In contrast, the extension of peri-urban areas ('suburban sprawl') is the more common form of urbanization in developed nations because their population is already heavily urbanized

(United Nations Population Division 2006). For example, in the United States the percentage population in urban centers has increased by only 10% from 1950 to 1990, but the area covered by cities has approximately doubled during the same period (Livernash and Rodenburg 1998). Both forms of urbanization involve slightly different environmental changes which are important to mosquito ecology, and thus will be considered in this chapter.

Since World War II, the world has experienced the (re)emergence of many infectious diseases. Some of these diseases are newly emerging in that their symptoms are distinct from any disease seen previously (e.g., Ebola virus, severe acute respiratory syndrome (SARS), and Nipah virus) (Daszak et al.,, 2000). However, most of these diseases are previously known and were thought to be under control, but are now re-emerging at rates not formally detected and in areas not previously affected (Barrett et al.,, 1998; Daszak et al.,, 2004; Weiss and McMichael 2004; Wilcox and Colwell 2005). The most pre-eminent diseases to re-emerge have been those vectored by mosquitoes (Diptera: Culcidae), such as malaria, filariasis, and arboviruses (arthropod borne viruses), which we will address in this chapter (Rosenberg and Jaeger 2000; Solomon and Mallewa 2001; Daszak et al.,, 2004; Mackenzie et al.,, 2004). Table 1 lists some of the important mosquito-borne diseases and their principal vectors recognized since World War II. The mechanisms and processes involved in the emergence and spread of vector-borne diseases are complex. However, human activities are frequently being identified as the primary cause (e.g., Aron and Patz 2001; Wilcox and Colwell 2005; Wilcox and Gubler 2005). In many parts of the world, the increase in vector-borne diseases has coincided with urbanization. These phenomena have led experts to postulate that the numerous ecological changes induced by urbanization cause an increase in mosquito-borne diseases (Gratz 1999; Patz et al.,, 2000; Daszak et al.,, 2001; Wilson 2001; Patz et al.,, 2004; Bradley and Altizer 2006).

Table 1. Re-emerging mosquito-borne diseases

Disease	Distribution	Mosquito vectors
Chikungunya	Africa, Asia	*Aedes (Stegomyia)*
Dengue fever / Dengue haemorrhagic fever	Africa, Americas, Asia	*Aedes (Stegomyia)*
Filariasis-bancroftian	Africa, Americas, Asia	*Culex* spp.
Japanese encephalitis	Asia	*Culex* spp.
Malaria	Global	*Anopheles* spp.
Rift Valley fever	Africa	Usually *Aedes* spp.
Venezuelan equine encephalitis	Americas	*Aedes* spp.
West Nile virus	North America	*Culex* and *Aedes* spp.

Two kinds of urbanization impacts have been highlighted as being the most important in increasing mosquito-borne diseases, and that are of interest in this chapter: (i) changes in host (human and animal) densities, and (ii) increases in vector population densities and distribution.

A localized increase in the human population is an inherent attribute of urbanization and can contribute to the spread of disease within regions through increased inter-host contact (Wilson 1995b). As we will discuss in this chapter, both human and animal movements can foster the spread of pathogens (an agent that causes disease, e.g., a bacterium, virus, or parasite) within populations and over considerable distances. However, increases in vector population densities usually have the greater role in determining if a mosquito-borne disease will become a medical problem (Gratz 1999). Even if a pathogen is introduced by a traveler and there is high host density, its establishment and spread is reliant upon the presence of a suitable vector in densities capable of transmitting and potentially sustaining an infection/outbreak. The different lifecycle stages of the mosquito are likely to be central to the two kinds of impacts described above. The terrestrial adult stage is where diseases are transmitted between hosts and thus instigate risks to human health. The aquatic larval stage is where the survival, development time, reproductive capacity, and population density are most likely to be affected by changes in environmental conditions. In examining both kinds of urbanization impacts, we identify relevant social, political, and economic factors that drive urbanization, and examine the environmental changes induced by urbanization that lead to increased disease risk.

We begin this chapter with a brief overview of the global public health threat that mosquitoes pose, focusing on the diseases that mosquitoes vector which present the greatest risk to humans. The known mosquitoes that constitute a public health threat represent a relatively small subset of all species, but they can vary considerably in their biology. We therefore consider the lifecycle of mosquitoes, evaluating which traits alter the responses of particular key taxa to certain environmental conditions. We continue this chapter by then reviewing how urbanization can increase mosquito-borne disease risk by (i) providing dense numbers of susceptible human hosts, (ii) encroaching on areas of high mosquito density, and (iii) increasing the density and spread of native and invasive species by creating favorable adult and larval habitats. Examples of mechanisms by which urbanization can increase mosquito-borne disease risk are summarized in Table 2. We refer to a species as 'invasive' if it is exotic and has established in a region, creating the potential of being a novel disease vector. By considering the invasion and spread of exotic vector species, we address another anthropogenic environmental change besides urbanization; in this case the spread of biota around the world through globalization. Throughout this chapter we will address how urbanization is often directly linked to other environmental changes, such as globalization, deforestation, agriculture, and water management to further amplify the threat from mosquito-borne diseases. Nonetheless, the focus of this chapter is on disease and mosquito dynamics within urban boundaries.

We then concentrate our review on the urbanization-mosquito scenario in New Zealand. This portion of our review will reveal how urbanization has specific synergistic effects with other environmental changes, both anthropogenic and natural, to present the likelihood of a mosquito-borne disease epidemic in New Zealand in the long term. In addressing the scenario in New Zealand, we hope to build on the ecosystem health paradigm proposed by other authors (Weinstein 1997; Forget and Lebel 2001; Parkes et al., 2003; Wilcox and Gubler 2005) to help understand the two-way link between human well-being and ecosystem health, and promote an ecological approach as a productive way to address the threat from mosquito-borne diseases in urban areas.

Table 2. Examples of mechanisms by which urbanization can increase mosquito-borne disease risk

Mechanism	Main effects on human, vector or pathogen biology and behavior	References
Urban populations susceptible to mosquito-borne diseases		
Increased inter-host contact	Human crowding increases the biting rate of mosquitoes, increasing the transmission of pathogens	Anderson and May (1991), Wilcox and Gubler (2005)
Invasion of novel pathogens and mosquitoes	Cities are hubs for travel and trade in an increasing connected world, and thus are the first entry points of exotic vectors and pathogens	Lounibos (2002), Juliano and Lounibos (2005)
	Resident populations are likely to have little immunity to novel pathogens and can sometimes be unfamiliar with behaviors to limit vector breeding and biting	Patz et al., (2000), Juliano and Lounibos (2005)
Migration of disease-susceptible humans	As people immigrate to cities for employment or as refugees they often are subjected to novel pathogens and have little immunity	Patz et al., (2000)
	Immigrating refugees often suffer from poor health making them especially susceptible to resident diseases, likely to encourage a new wave of disease, or introduce a novel disease	Pirages et al., (2001)
Urban encroachment on mosquito habitats		
Encroachment on agricultural land and native habitats	Exposes humans to greater numbers of vectors and existing disease cycles	Ljumba and Lindsay (2001), Russell et al., (2002), Old and Deane (2005)
	In developing nations, immigrant workers with little disease immunity frequently live in settlements bordering agricultural areas	Ljumba et al., (2002)
Urban agriculture	In developing nations, agriculture often occurs within urban landscapes, thus placing humans and potential vector mosquitoes within close proximity	Afrane et al., (2004), Matthys et al., (2006)
Environmental changes that create favorable mosquito habitat		
Population level effects on urban disease vectors		
Creation of new breeding sites	Human activities create numerous types of favorable breeding sites that increase the production of adult mosquitoes	Regis et al., (1995), Kay et al., (2000), Vezzani (2007)

Table 2. Continued

Mechanism	Main effects on human, vector or pathogen biology and behavior	References
Community level effects on urban disease vectors		
Altered conditions of existing habitats	Increased ambient air temperatures ('heat island effect') can lead to greater adult survival and vectorial competence	Lindsay and Birley (1996)
	Changes to the physical and chemical characteristics of breeding sites can favor the production of adult mosquitoes	Coluzzi (1994), Lindblade et al., (2000)
Altered host-vector interactions	Non-competent reservoir hosts that dilute rates of disease transmission ('dilution effect') are removed, increasing human infection risk	Service (1991), Ezenwa et al., (2006), Kilpatrick et al., (2006)
Favorable community dynamics in breeding containers	Ecological niches of vector species expand as predators, competitors, and parasites are reduced or eliminated ('ecological release')	Juliano and Lounibos (2005), Wilcox and Gubler (2005)

Public Health Threat from Mosquitoes

The incidence of mosquito-borne diseases is cited by a number of international organizations as one of the most serious public health problems facing the 21st century (WHO 1982; WRI 1998; Rosenberg and Jaeger 2000). Many major mosquito-borne diseases (including varieties of malaria, dengue, and filariasis) are now occurring over wider geographical and ecological areas, and are affecting more people than ever before (Table 1). In terms of mortality and morbidity, malaria and dengue have the greatest public health burden (Weiss and McMichael 2004).

After a decline in the 1950's and 60's, malaria, the world's most deadly mosquito-borne disease, has resurged in many parts of the world to claim 1–3 million lives and cause 300–500 million acute illnesses globally per year (WHO 1996). In Africa alone, malaria is the third most frequent cause of death, behind only HIV/AIDS and lower respiratory syndrome (WHO 2002). In the last 20 years, dengue, and its most severe form, dengue hemorrhagic fever (DHF), has increased at even a faster rate as malaria. Today, new infections of dengue/DHF are reported at 20-50 million cases globally, per year (Wilcox and Colwell 2005), up from an average of about 30,000 cases in the 1960s (WHO 1996). Dengue/DHF has spread to 101 countries and is present in almost all of Southeast Asia, where it is one of the leading causes of pediatric hospitalization and mortality (Halstead 1993).

The introduction and spread of a deadly arboviral disease even in a nation with the most modern health services is no better demonstrated by the spread of West Nile virus across the USA. After first being reported in the New York City area on the East coast of the USA in

1999, West Nile virus has swept across all but two states (Tyler 2004). From 1999 to 2004, West Nile infected over 16,500 people and its most serious manifestation, fatal encephalitis, caused 666 human deaths in the USA (CDC 2007). As we will discuss below, the spread of mosquito-borne diseases, such as these, have been attributed to the influence of urbanization and associated environmental changes on human demographics, vector ecology, and the invasion of mosquitoes into new regions.

Mosquito Lifecycle and Vector Capacity

Mosquitoes belong to the dipteran family Culicidae; comprising over 3,500 species, which are divided into 38 known genera (Service 2000). Only a small proportion of mosquito species are known vectors for human disease but many more can exist in such high densities that they present a considerable nuisance (Service 1995). The most medically important genera are *Anopheles*, *Culex*, and *Aedes*. *Anopheles* mosquitoes are vectors for malaria and are grouped within their own subfamily of Anophelinae. *Culex* and *Aedes* are vectors of a range of diseases and are in the subfamily Culicinae.

All mosquitoes share a similar model of development (Figure 1). Female mosquitoes can lay up to 500 eggs at one time. In the case of *Culex*, the eggs are fixed together in rafts, which float on the surface of the water. The construction of egg rafts allows *Culex* mosquitoes to lay relatively large batches of eggs in a short time. *Anopheles* and *Aedes*, as well as many other genera, do not make egg rafts, but rather lay their eggs singly. Single *Anopheles* eggs float on the water surface. Most species of *Aedes* lay their eggs on damp soil substrates or in container habitats (e.g., plant axils, used tire casings, guttering, vases, and pots) above the water surface that are later inundated. Eggs generally hatch into larvae within 48 hours. However, mosquitoes that lay their eggs away from the water surface have water-proof shells and show varying degrees of tolerance to desiccation. Some eggs last for only a few weeks while others can last for up to two years. Desiccation-resistant eggs are strongly associated with becoming an introduced species, presumably because they increase the probability of successful transport (Juliano and Lounibos 2005).

Ideal water bodies are usually shallow, lentic, and warm. As we will discuss in the following sections, urban landscapes provide a large number of suitable habitats for mosquitoes, particularly species that breed in container habitats that easily fill with rain-water or are artificially watered. Larvae hatch and develop through four instars. Mosquito larvae are essentially feeding machines, with the vast majority of species filtering bacteria, detritus, and algae within the water (Merritt et al., 1992). Larvae can take from a couple of days to several weeks to develop, depending on a range of environmental factors, including temperature, nutrient level, competition, predation, and parasitism. While sharing essentially the same pattern of biological development, mosquito species can differ considerably in their physiological tolerances, particularly during the aquatic larval stages (Clements 2000a).

After a brief pupal stage, winged adults emerge. Adults usually mate soon after hatching and feed on nectar, but females of all but a few species require a bloodmeal to provide proteins and nutrients for egg development and the continuation of the lifecycle. Females often blood-feed from multiple hosts because they either get insufficient blood required for oviposition from one host or survive long enough to lay multiple batches of eggs. The blood-feeding habit of female adults renders them prone to acquiring pathogens and transmitting diseases when they feed on an infected host and then bite a non-immune human.

Because much of their lifecycle is outside the human host, vector-borne pathogens are particularly sensitive to the environment. Environmental changes can impact pathogens directly by altering their incubation time and indirectly by affecting the distribution, abundance, longevity, development time, and biting behavior of vectors (Wilson 1995a).

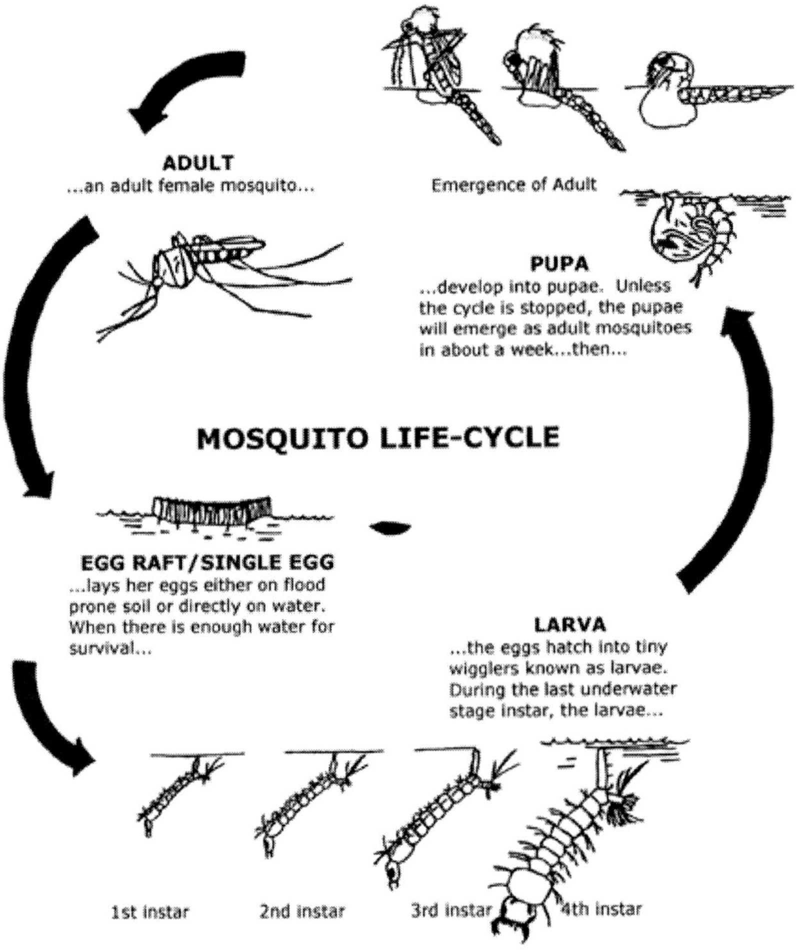

Figure 1. The generic developmental lifecycle of mosquitoes (Diptera: Culicidae).

The contribution of environmental variables to the spread of vector-borne diseases is summarized in a measure called the vector capacity (Wilson 2001):

$$VC = (ma)\ (P/F)e^{-n/E}E$$

where ma is the number of bloodmeals on humans taken by the vector population per person per day, P is the proportion of bloodmeals taken by a vector on humans, F is the time interval between bloodmeals taken by a vector (days), e is the base of natural logarithm (~2.71828), n = time for development of the infectious agent (days), and E = vector longevity (days). Later in this chapter, we will address how urbanization can dramatically increase the incidence of mosquito-borne diseases by a number of interacting environmental changes that all alter the ecological balance and context within which vectors and their pathogens breed, develop, and transmit disease.

Urbanization and Human Health

Since the Industrial Revolution in Europe and North America 500 years ago, urbanization has been associated with an overall improvement of the quality of life in developed nations (Pirages et al., 2001). Urbanization has fostered advancements in medical science, technology, and infrastructure, and although subject to regional variation, these advancements have generally contributed to improvements in living standards, decreased mortality rates, and the reduction of many major infectious diseases, including the eradication of smallpox and dramatic decline in polio (Barrett et al., 1998). Gradually, these developments in health care diffused to less-developed regions (World Bank 1993), and in particular, to major urban centers in these regions.

The benefits of urban development, however, have often been counteracted by stresses to the environment and human health. Huge environmental changes have resulted as a consequence of a scale shift in numbers of people, density of settlement, and the intensity of new manufacturing processes that all accompany the growth of cities. Urbanization led to the deterioration in the health of the central districts of western cities during the Industrial Revolution as these cities contended with foul air, unclean water, and inefficient sanitation (Pirages et al., 2001). Today, cities in developing nations are experiencing the same problems (Meyerson et al., 2007). In addition, the re-emergence of many infectious diseases since World War II, including those vectored by mosquitoes, has been attributed to environmental change, and mark what some authors consider a *third epidemiological transition* (McMichael and Haines 1997; Barrett et al., 1998; Weiss and McMichael 2004). The benefits and environmental costs of urbanization have not been uniform in the global context. Although having a lower proportion of their population in cities, nations in the early stages of development such as China, India, and Brazil are among the most populous in the world, have more than double the number of urban dwellers as developed nations (2.3 billion vs. 0.9 billion) (United Nations Population Division 2006), and thus typically undergo the most substantial environmental changes to transform their economies and improve living standards (Meyerson et al., 2007).

Urban Populations Susceptible to Mosquito-Borne Diseases

Increased Inter-Host Contact in Cities

Well established principles of theoretical population biology, applied via mathematical epidemiology, show that the incidence of infectious diseases increases with human density (Anderson and May 1991). Moreover, depending on the type of pathogen, vector species, and host population, mathematical theory has also demonstrated the existence of host-population density-dependent thresholds that determine if diseases will (re)emerge and be sustained in a population (May and Anderson 1979). Typically vector-borne diseases have lower thresholds than directly transmitted pathogens, thus allowing a greater likelihood of their establishment even in relatively small urban settlements (Anderson and May 1991). Thus, the coalescence of scattered settlements into fewer large industrialized cities has led to the abrupt transition of mosquito-borne diseases from non-persistence to epidemic behavior (Wilcox and Gubler 2005).

Since vector mosquitoes acquire pathogens while feeding on an infected host and transmit the agents during a later bloodmeal, the effect of crowding is not as straightforward as for directly transmitted agents. In particular, the infectious status and immune status of human hosts also affects the transmission and expression of mosquito-borne diseases (Anderson and May 1991), and social, economic, and political processes can modify the effects of urbanization on disease incidence and likelihood of outbreaks (Sala et al., 2000). These effects include the introduction of novel diseases and vectors between regions and large-scale migration of susceptible humans.

Invasions of Pathogens and Mosquitoes into Cities

Because cities are central to trade, commerce, and travel, they are usually the first point of entry for an exotic mosquito or pathogen into a new region. The risk of entry and possible establishment of exotic vectors and pathogens has grown in concert with increased globalization (Lounibos 2002), which is reflected by record volumes of international travel and trade occurring year after year since the 1950s (The World Bank 1997; The World Bank 2007). Novel pathogens can typically spread more quickly than normal once reaching a new city because the host population will likely not have developed protective (sometimes called 'herd') immunity (Juliano and Lounibos 2005) or be familiar with behaviors that would limit their availability as hosts to vectors, e.g., using bed nets and remaining indoors during peak biting hours (Patz et al., 2000).

Mosquitoes, can be spread by land vehicles, airplanes, and shipping (Laird 1995). Species with immature stages that occur in artificial container habitats and have desiccation-resistant eggs have been responsible for the most notable invasions and range expansions at the regional and global scales (Cook et al., 2002; Lounibos 2002) because artificial containers are commonly used during transport and in urban settings (Laird et al., 1994; O'Meara 1995). Pathogens can be introduced with a mosquito vector or independent of a vector, e.g., via movement of an infected host. In other instances, invasive mosquitoes have acquired an independently introduced pathogen sometime after its establishment (Juliano and Lounibos 2005).

When exotic mosquitoes invade a city they may vector an existing disease. Occasionally the exotic mosquito may be a less efficient vector than the resident species, but it can also be a more efficient vector and thus increase disease incidence (Juliano and Lounibos 2005). Some invasive vectors may expand into new areas by filling empty ecological niches, such as by occupying unused habitats or exploiting unused resources (Williamson 1996). Empty niches may be created through the proliferation of new breeding sites, alterations to the physical and chemical properties of existing habitats, or the loss of resident biodiversity. Alternatively, if resources are limiting, invasive vectors may replace resident species if they are competitively superior or more efficient at escaping predation and parasitism (Juliano and Lounibos 2005).

Migration of Disease-Susceptible Humans into Cities

Large-scale human migration potentially facilitates greater disease spread than single infected travelers. As rural people move into urban areas for employment they are potentially exposed to a suite of novel diseases or they may introduce novel pathogens into the local population (Patz et al., 2000). In the last three decades, a large proportion of migrants have been refugees generated by resurgence in ethnic conflict in Africa, Asia, and the Middle East (Pirages et al., 2001). These populations typically form or are forced into crowded refugee camps or move to extremely localized areas within resident urban areas. Refugees often suffer from poor health mediated through the effects of malnutrition and inadequate sanitation. Hence, they are especially susceptible to resident infectious diseases already circulating in a city, encouraging a new wave of disease, or introducing a novel disease (Pirages et al., 2001).

Urban Encroachment on Mosquito Habitats

In addition to increasing human population density, urbanization can increase the risk of infection by the expansion of urban areas into areas of high mosquito density. These areas can include a range of landscapes, but the most common in respect to mosquito-borne disease are agricultural land and native habitats, e.g., wetlands and forest margins.

Extensively irrigated crops, the most common being rice paddies, involve creating large areas of standing ground water suitable for extensive mosquito breeding (Lacey and Lacey 1990; Ljumba and Lindsay 2001). Outbreaks of malaria (Ljumba et al., 2002), Japanese encephalitis (Mackenzie et al., 2004), and lymphatic filariasis (Service 1991) have commonly occurred in urban areas that adjoin rice paddies. Often these settlements have high densities of immigrant workers that may have little natural disease immunity and thus further increase disease incidence (Ljumba et al., 2002). In addition, in developing nations, agriculture often occurs within urban landscapes (called 'urban agriculture'), which further facilitates the overlap between many vector species and human populations (Afrane et al., 2004; Matthys et al., 2006). Agricultural practices can provide additional sources of breeding habitat, such as water reservoirs, stock drinking troughs, and rain-filled ditches (Service 1991; Norris 2004). Increased densities of farmed animals can alter the transmission cycles of pathogens (Mackenzie et al., 2004). Livestock may reduce the rate at which vectors feed on humans ('dilution effect', discussed in sections later in this chapter) (Sota and Mogi 1989), but they can also increase disease incidence by increasing vector density and maintaining the virus in the environment (Patz et al., 2000).

The encroachment of urban areas into native habitats can facilitate both naturally existing disease cycles and the transmission of new diseases to the human population who are susceptible to these novel pathogens. For many diseases, non-human animals are the primary host and human infections are often incidental. However, despite being incidentally infected, humans can suffer severe health effects and mortality from such diseases. These diseases can present a substantial public health burden when generalist vector species act as bridges between primary and human hosts by feeding sufficiently abundantly on both. Such a scenario appears to have facilitated the introduction of yellow fever into many urban areas in both Africa and the Americas. Rapidly expanding settlements that fragment jungle habitats have allowed domestic *Aedes* species to feed on both non-human primates and humans (Petersen and Marfin 2005). In Australia, residential and industrial expansion into estuarine areas have changed the drainage patterns in mangrove and saltmarsh habitats increasing the distribution of *Aedes vigilax* and *Culex sitiens,* resulting in outbreaks of Ross River virus (Russell et al., 2002). Similar urban expansion in Australia into habitats containing native mammals, including marsupials, has also been suggested as possible causes for the increased incidence of Ross River virus disease in urban settings (Old and Deane 2005). Greater encroachment of urban areas into native habitats is expected to increase the incidence of infection by altering the feeding behavior of many species from zoophyllic (animal-to-animal) to anthrophyllic (human-to-human) orientation and thus further increase the disease burden (Anderson and May 1991; Patz et al., 2000; Lounibos 2002).

Urbanization, Environmental Change and Mosquito Ecology

Globally, humans have converted approximately 4% of the earth's land area into urban landscapes (WRI 2007), and by 2030, this is expected to double (United Nations Population Division 2006). Urban driven land conversion has become more pronounced over the last century because of steep increases in the global population, social inequalities, and per capita resource consumption (UNEP 1999; WRI 2000). Urbanization has been shown to decrease biodiversity and natural habitat, increase densities of human structures (e.g., roads, buildings), and shift microclimates (e.g., heat island effects).

Urbanization has pervasive economic and ecological links with other land-use types. Continued growth of urban areas frequently dictates the expansion or development of more intensive land-uses for agriculture, energy, waste, and water management. The highly concentrated urban populations also provide the market, and hence economic justification, for other environmental changes besides agriculture, such as mining, commercial fishing, and intentional introductions of exotic biota (Wackernagel and Rees 1996). Folke et al., (1997) estimated that the cities of Baltic Europe had an impact or *ecological footprint* on areas of forest, agricultural, marine, and wetland ecosystems that is 565-1130 times larger than the area of the cities themselves. These direct and indirect environmental impacts of urbanization can alter the ecological balance and context within which vectors and pathogens reproduce, develop, and transmit disease, and, ultimately, often lead to higher incidence of mosquito-borne diseases (Patz et al., 2000). In the following sections, we will address the cascade of ecological disruptions that occur within urban boundaries that can alter the densities and spread of native and exotic mosquitoes. We divide the ecological disruptions into those that: (i) directly impact the population size of one or more species by altering the available ecological niche space, and (ii) involve community-level species interactions as a result of altered or newly created ecological conditions.

Population Level Effects on Urban Disease Vectors

Mosquito vector densities are widely considered the foremost factor in determining vector borne disease burden. Theoretical population biology has demonstrated that pathogen 'reproductive rate' increases with the square of the vector population density (Anderson and May 1991). This effect is well illustrated by the dengue pandemic. Dengue occurs in the world almost wherever *A. aegypti* is present, including in many Central and South American countries where it renewed transmission in the 1990s after *A. aegypti* re-invaded and spread (Fanning et al., 1997; Gratz 1999). Dengue/DHF has also established and spread in cities without *A. aegypti* when *A. albopictus* is in sufficiently high densities (Gratz 1999). Thus, diseases can (re)emerge extremely quickly when vectors increase their density and expand their distribution, even if initiated by a single infection event.

Urbanization alters the basic physical properties of the ecosystem, such as the predominant vegetation, hydrology, soil structure, and topography. These changes in turn alter the complex chemical and biotic processes that typified the original ecosystem (Patz et al., 2000). This change often leads to the elimination or severe disruption of the ecological niches of resident species, whilst new ecological niches may be formed. There are two main mechanisms by which urbanization can alter ecological niches of mosquitoes to increase vector densities. The first mechanism is the expansion of existing niches though the proliferation of favorable breeding sites of vectors. The second mechanism is via the creation of new niches by making existing habitats more favorable for particular vector species. Both mechanisms can proceed with little inter-specific interaction, and has probably been the main way by which *A. aegypti* has invaded and spread into non-native regions (Juliano and Lounibos 2005).

Creating New Breeding Sites

Artificial containers, such as tire casings, guttering, water-storage tanks, and plant-pots are commonly used in human activities and are thus inherent entities of urban landscapes. When artificial containers collect water from intended watering or rainfall they provide ideal habitat for numerous mosquito species. Cemeteries have been known to support extremely high densities of container breeding species in both developed and undeveloped nations (Juliano et al., 2004; Vezzani 2007). For example, studies in South America have estimated that cemeteries may produce tens of millions of mosquitoes per year (Vezzani 2007), while the output of *A. aegypti* from a relatively small (3.5 hectare) cemetery in Venezuela was estimated to be 3,000 per day (Abe et al., 2005). Complex subterranean stormwater drains also provide ideal habitat for *Culex* and *Aedes* mosquitoes, particularly in regions and during seasons when surface water is scarce. Kay et al., (2000) have shown that stormwater drains are suitable for a range of vector species in Australia, and may account for as much as 78% of the resident urban mosquito population in the dry season. Perhaps the greatest vector densities are observed in cities of developing tropical countries where urbanization is poorly planned, an example being rapid urbanization in Brazil which supported the invasion of *Anopheles gambiae* in the 1930s (Soper and Wilson 1943). Often piped water supplies are inadequate in cities of developing nations and residents have to store rainwater and well water that can also provide mosquito habitat (Gratz 1997). Continued urban sprawl in the Americas since World

War II has exceeded the capability of sanitation services to collect disused containers that support extremely high densities of *A. aegypti* (Gratz 1997). Other examples of human activities that create mosquito breeding sites include road construction which allows rainwater to pool in undulations in the ground (Vittor et al., 2006), while open waste water systems (e.g., cess pits, septic tanks) can also substantially increase densities of mosquitoes suited to organically rich breeding sites, such as *Culex quinquefasiatus* in urban cess pits in Brazil (Regis et al., 1995).

Altering the Conditions of Existing Habitats

Because species vary in their environmental preferences and physiological tolerances, changes to the physical and chemical environment are likely to affect species differently. Thus, depending on the relevant environmental impacts in any given circumstance, the population of some vector species will be favored by urbanization, while others will be negatively affected and may even be eliminated. This scenario can often lead to an increase in the abundance of vectors not previously considered important in regions (Rubio-Palis and Zimmerman 1997). For instance, the desalinization of city water intended for drinking and agriculture has had the unintended consequence of increasing the transmission of malaria by altering the locally predominant species. Desalinization combined with poor drainage and piping has favored the replacement of a less effective malaria vector, *Anopheles melas,* with the more efficient vector *A. gambia*e (Coluzzi 1994).

The creation of new ecological niches is also often linked to changes to the shading and nutrient regimes of the environment, and has led to increased malarial transmission rates in both the Americas and Africa (Antonio-Nkondjio et al., 2005). Urbanization frequently results in the clearing of resident vegetation. This can increase local ambient and water temperatures as well as cause erosion and sedimentation that alter the chemical composition of breeding sites (Patz et al., 2000; Vittor et al., 2006). The creation of turbid, sunlit pools as a result of deforestation and settlement has favored the disproportionate proliferation of *A. gambiae* compared to other less efficient malarial vectors in Africa (Sattler et al., 2005; Matthys et al., 2006).

Besides favoring a more efficient malarial vector, urban induced deforestation and nutrient increases can also lead to 'hyper-abundant' mosquito densities due to greater larval growth rates (below an upper lethal threshold) (Clements 2000a; Lindblade et al., 2000) and microbial productivity on which larvae feed (Wetzel 1995; Wetzel et al., 1995), further amplifying malaria transmission.

Besides making aquatic larval sites more favorable for mosquito production, urbanization also alters adult habitats and promotes greater adult survival and vectorial competence. For instance, developed urban landscapes usually have higher ambient air temperatures than neighboring rural areas because human-made buildings generally provide radiant heat energy (i.e., heat island effect) (Frankie and Ehler 1978; McIntyre 2000). Increases in temperatures from buildings are likely to lead to a reduced incubation time for pathogens and greater transmission rates (Lindsay and Birley 1996).

The proximity of people to mosquito favorable breeding sites directly affects their likelihood of being bitten by mosquitoes and infected by pathogens. The more favorable larval and adult habitats are to vectors, the higher the risk of pathogen transmission. In a study by Thompson and Begtrup (1997) in Mozambique, the risk of malaria was 6.2 times higher for individuals living less than 200 meters from *Anopheles* breeding sites than for individuals living 500 meters or more away. In Australia, dengue virus exposure from *Aedes* mosquitoes has been shown to be 2.5 times higher for people living within 160 meters of an access point to subterranean stormwater drains (Russell et al., 2002).

In another example from Australia, during a Ross River virus outbreak in the Mandurah region of south-west Western Australia the disease incidence was 10 times greater in residents within 3 kilometers of saltmarsh mosquito habitats; over 64% of cases resided within 1 kilometer of saltmarsh habitats, 98.8% of cases resided within 3 kilometers, and all 162 cases were located within 5 kilometers (Russell 2002). Thus, greater densities of favorable breeding habitats and population crowding from urbanization can potentially initiate epidemics of mosquito-borne diseases.

Community Level Effects on Urban Disease Vectors

When habitats or resources are limiting, biological communities are largely structured by inter-specific interactions, and the distribution and abundance of each species can potentially affect another (Begon et al., 1990). The loss of resident diversity is thought to make communities more susceptible to further disturbance, collapse, domination by a few species, and the invasion of exotic species (Peterson et al., 1998; Simberloff and Von Holle 1999). Since World War II, urbanization has coincided with reductions in biodiversity and the simplification of communities (Sala et al., 2000; Daszak et al., 2001; Pauchard et al., 2006; Pongsiri and Roman 2007). Recently, ecologists have begun to explore mechanisms by which urbanization affects biodiversity, community dynamics, and disease (Faeth et al., 2005; Pongsiri and Roman 2007). The density and condition of mosquitoes and, ultimately, the infection risk of disease is likely to be highly affected by predation, competition, and parasitism in larval breeding habitats and host-vector dynamics during the adult stage (Juliano and Lounibos 2005).

Host-Vector Interactions

A phenomena known as the 'dilution effect' predicts that infection rates among vectors, and ultimately human infection risk, will be lower in highly diverse host communities where non-competent reservoir hosts dilute rates of disease transmission between vectors (Schmidt and Ostfeld 2001). Lyme disease, caused by the bacterium *Borrelia burgdorferi,* which is transmitted by ticks that feed on mammals, is the best studied example of the dilution effect. Studies have shown that suburban environments in northeastern USA have low mammalian biodiversity and high abundance of the most efficient host species (white-footed mice), creating a simplified community that promotes high infection rates in both ticks and humans (Ostfeld and Keesing 2000; LoGiudice et al., 2003). A similar scenario occurs in peri-urban settlements in Africa, Asia, and the Americas, where human biting rates and malarial incidence increases in areas where urbanization and rice cultivation displace livestock and reduce biodiversity (Gilglioli 1963; Service 1991). It is thought that the dilution effect plays a similar role in other mosquito-borne pathogens, including West Nile virus in North America when urbanization reduces the number of non-passerine bird species (Ezenwa et al., 2006; Kilpatrick et al., 2006).

Community Dynamics in Breeding Containers

The reduction of biodiversity can also contribute to a phenomenon called 'ecological release' in the remaining species, whose ecological niche expands as predators, competitors, and parasites are reduced or eliminated (Begon et al., 1990). If the remaining species are vectors for disease, an increase in infection rates is likely to occur as their population abundance expands (Wilcox and Gubler 2005). If exotic vector species invade a non-native area they may escape natural inter-specific population regulators and initiate a disease epidemic. However, species-rich communities are more likely to have competitors or predators that act as barriers to potential invaders even if their fauna has not co-evolved with the invader (Levine and D'Antonio 1999). These community dynamics help to explain why mosquito-borne diseases often proliferate in urban landscapes whose aquatic communities have been disturbed (Wilcox and Gubler 2005). For instance, urbanization induced ecological changes that coincide with the replacement of the dominant malarial vectors (discussed in sections above) are likely to be partly driven by resource competition, with the new vector having competitive superiority in the new conditions (Schneider et al., 2000; Depinay et al., 2004). However, to our knowledge, this has not been definitively tested. In the case of *A. gambiae* invasion into the Americas in the 1930s and continued spread throughout parts of Africa, intra-guild predation (i.e., the predation of later instars of one species on early instars on another species) may facilitate the more efficient malarial vector in urban areas (Koenraadt and Takken 2003; Juliano and Lounibos 2005). Evidence for reduced biotic resistance of urban landscapes to species invasion (Case 1990) and disease proliferation (Bradley and Altizer 2006) is rarely described and inconclusive and is likely to vary considerably between systems.

Currently, much of our understanding on the effects of urbanization on mosquito communities is based on recent invasions of North America and South America by *Aedes albopictus*. *Aedes albopictus* is considered medically important because it is a vector of arboviruses, including dengue, La Crosse encephalitis, and West Nile virus (Ibañez-Berñal et al., 1997; Gerhardt et al., 2001; Turell et al., 2005). Its invasion and spread in the Americas appears to be influenced by urban induced changes to resident competitors and predators. For instance, the presence of native *Wyeomyia* species in Florida appears to limit the abundance of *A. albopictus* in cities, presumably because the native is the superior competitor in bromeliad plants where both species co-occur (Lounibos 1985). While bromeliads are commonly planted in suburban areas, they are unlikely to be found in the central core of crowded cities that are dominated by industrial and commercial zones. In their absence, the generalist *A. albopictus* is likely to breed in artificial containers where the competitively superior *Wyeomyia* species do not breed. This may lead to higher region-wide population densities of *A. albopictus* and increase potential disease risk.

The establishment and spread of *A. albopictus* is also likely to be facilitated by urban-induced reduction of invertebrate predators' *Toxorhychites rutilus* and *Corethrella appendiculata.* Both predators preferentially breed in natural forests, particularly treehole habitats (Lounibos 1985); therefore, as urban landscapes have expanded, deforestation has led to their decline. Laboratory experiments show that *A. albopictus* is highly vulnerable to both predators (Lounibos 1985). Field studies showing the failure of *A. albopictus* to spread into forested areas despite its competitive superiority over resident mosquito species and domination in neighboring urban landscapes, suggests a prominent role of these native predators in limiting *A. albopictus* abundance (Lounibos 2002).

Interestingly, the invasion of *A. albopictus* in North America may also illustrate more subtle mechanisms by which urbanization may modify inter-specific interactions. *Aedes aegypti* is the most common species to co-occur with *A. albopictus* in larval habitats in the southeastern United States. *Aedes aegypti* has declined, often to local extinction, in association with the spread of *A. albopictus* in the Southeast (O'Meara 1995; Juliano et al., 2004). Evidence from field and laboratory experiments indicates this decline is probably because of competitive displacement (Juliano 1998; Juliano et al., 2004). Seemingly paradoxically, *A. aegypti* coexists with *A. albopictus,* and is often the dominant species, in many urban cities in the southeastern USA (Juliano et al., 2004). Laboratory and field data suggest that that urban-induced microclimatic shifts towards warmer, dryer habitats and resultant egg mortality differentially affects *A. albopictus* and reverses the competitive advantage to favor *A. aegypti* (Juliano et al., 2002; Costanzo et al., 2005). Metapopulation dynamics have been hypothesized as another mechanism by which urbanization may facilitate the persistence of *A. aegypti* in Florida, as the species may be able to escape competition from *A. albopictus* by being a superior colonizer between 'islands' of extremely favorable breeding habitat (e.g., cemeteries, tire dumps) (Juliano et al., 2004). Because *A. aegypti* is a more efficient vector for dengue/DHF and yellow fever, its persistence poses a risk of these diseases in urban areas. Although dengue/DHF is not currently present in North America, similar inter-specific interactions between *A. aegypti* and *A. albopictus* are likely to be occurring in other parts of the world where the two species and disease co-occur.

New Zealand Scenario

Urbanization in New Zealand

New Zealand is a remote cluster of islands in the South Pacific east of Australia, lying between 34° and 46° south longitude. New Zealand consists of three main islands (North, South, and Stewart) that are orientated north-south and consist of c.27 million hectares, or about the same size of the British Isles or Japan. Given its isolation, New Zealand was first settled by humans only relatively recently, c.800 years or less than 30 generations ago. New Zealand's first inhabitants, Polynesian-speaking Māori, formed relatively small settlements ranging from 300 to 500 households in areas that were rich in natural resources (e.g., coastal fisheries, rivers) (MfE 1997). As their population grew, Māori increasingly changed their surrounding environment, primarily in the form of deforestation from fires that were for agriculture, hunting, or by accident. Endemic mosquitoes appear to have been important nuisance insects to Māori. Māori gave mosquitoes the names of naeroa or waeroa (Reed 1963), and commented on their annoying biting to early Europeans (Polack 1838). Māori settlements undoubtedly provided relatively dense clusters of hosts on which mosquitoes could feed (Reed 1963), however, it is not clear whether human activities greatly altered mosquito breeding habitats at this early time.

The most extensive anthropogenic environmental changes have occurred in New Zealand within the last 150 years after European settlement and subsequent large-scale (mainly British) immigration. New Zealand's native forests have been reduced from covering approximately 85% of the total land area to 23%, with about half of all deforestation occurring since European settlement and agriculturalism became prevalent (MfE 1997). Open pastureland is now New Zealand's most extensive land type constituting 50% of the country's total land area (MfE 1997), nearly two times higher than for world estimates (26%, WRI 2000). However, if other farming practices are included, such as horticulture, this figure increases to 58% (Statistics New Zealand 2004).

Steady urbanization has accompanied increases in immigration and agriculturalism. Following World War II, increased industrialization drew large numbers of rural workers into New Zealand's cities as laborers and factory workers, ushering in a period of particularly rapid urbanization. Today, approximately 86% of the population live in urban areas (MfE 2006), and in 1997, urban landscapes covered approximately 3% of the total land area of New Zealand (MfE 1997). The high percentage of urban dwellers in New Zealand makes the country one of the most urbanized countries in the world. Although the urban population in New Zealand only grew by 30% from 1969 to 1997, the amount of land classified as urban has almost trebled, increasing on average 4% per annum (WRI 2000). Much of the remaining native ecosystems are restricted to mountainous and upland areas. Lowland areas that are most amendable to agricultural development have been almost totally modified to create a patchwork of isolated fragments of varying land cover (DoC and MfE 2000).

The replacement of biodiverse native ecosystems with agricultural monocultures and urban landscapes has facilitated the decline of native biodiversity and the invasion, establishment, and spread of exotic species (Cook et al., 2002). The manifold effects of these disruptions have severely altered ecological patterns and processes of the original environment. For instance, the exotic Australian brushtail possum (*Trichosurus vulpecula*), which numbers as many as 70 million (Parliamentary Commissioner for the Environment 2000), thrive in human-modified landscapes but also invade neighboring native areas where they directly feed on native flora and fauna (Cowan 1990).

Two vector-borne pathogens leading to human infection have recently emerged in New Zealand, and ecosystem disruption, through urbanization, is proposed as a driving mechanism. Exotic rats (*Rattus* spp.) and introduced domestic cats (*Felis domesticus*) are the main reservoir hosts for pathogenic agents (*Rickettsia typhi* and *Bartonella henselae*, respectively) that cause disease in humans (Kelly-Hope et al., 2005). The known number of human infections of both diseases is still low, but rates of Rickettsial disease have increased (ESR 2007). The actual incidence of both diseases is probably considerably higher (Kelly-Hope et al., 2005). However, more importantly, these diseases represent examples of how introduced vector borne diseases have recently emerged in New Zealand.

New Zealand's Mosquito Fauna

As of 2007, New Zealand has 12 endemic and four exotic mosquito species, of which, one exotic is subject to an ongoing government eradication program. All four exotic species are known vectors for human diseases in other parts of the world, including a wide range of arboviruses including West Nile virus, Kunjin and Sinbis viruses, Barmah Forest virus, dengue and Japanese encephalitis viruses, and Ross River virus (Russell 1995). Of particular concern is the presence of the Australian exotic species *Aedes camptorhynchus*, which is an efficient vector of Ross River virus. However, this species is not a container breeder and thus may not maintain or spread a disease outbreak in an urban setting. In an urban environment the species of concern are the Australian exotic *Aedes notoscriptus* and the pantropical/subtropical *Culex quinquefasciatus*. This is because of their potential as disease vectors, container-breeding behavior, and emergence as the most common human-biting species in New Zealand's urban centers (Laird and Easton 1994).

Elsewhere in the world, *Aedes notoscriptus* is a potential vector of Barmah Forest virus (Watson and Kay 1999), a laboratory vector of Rift Valley Fever (Turell and Kay 1998), dengue and Japanese encephalitis viruses (van den Hurk et al., 2003), and has been implicated as a vector of Ross River virus in Australia in urban areas (Doggett and Russell 1997; Watson 1998). Also overseas, *Culex quinquefasciatus* is an efficient vector of West Nile virus (Tyler 2001), St. Louis encephalitis, lymphatic filariasis (*Wuchereria bancrofti*) (Day 2001), Japanese encephalitis (De and Chandra 1994), Murray Valley encephalitis, Kunjin and Sindbis viruses (Russell 1995), and a suggested carrier of Ross River virus in Australian cities (McLean 1953; Lindsay et al., 1996).

New Zealand's native mosquito fauna is unusually depauperate compared to other countries of similar size and latitude (e.g., England, 32 species; Japan, 67, species) (Holder et al., 1999) and appear to underutilize available habitats (Belkin 1968; Laird 1995). Little is also known about their vector capability because no studies examining their competence have been undertaken. However, the recent isolation of *Plasmodium relictum* from the endemic species *Culex pervigilans,* which is estimated to be New Zealand's most common and widespread species (Belkin 1968; Laird 1990; Laird 1995), has implicated this mosquito species as playing a role in avian malaria transmission in New Zealand (Massey et al., 2007). In addition, the avarian Whataroa virus (a Sinbis-type arbovirus) has been isolated from *C. pervigilans* and another endemic species, *Culiseta tonnoiri* (Maguire et al., 1967; Miles 1973). A similar virus (Ockelbo) in Sweden causes outbreaks of polyarthritis (Lundstrom et al., 1991), suggesting that New Zealand is not too cold to be susceptible to outbreaks of arboviruses.

New Zealand's endemic species likely pose a lower risk of becoming prominent vectors of introduced pathogens. Although anecdotal evidence suggests that they feed off humans and can become nuisances, they are primarily bird-feeders (Derraik and Snell 2004). However, their bird-human biting behavior and the close relatedness (same subgenus) of some species to overseas and resident exotics that are known vectors of human diseases, may make them ideal bridge vectors for a range of zoonotic pathogens, including West Nile virus (Spurr and Sandlant 2004).

Urbanization and Mosquito-Borne Disease Risk

There has never been a confirmed indigenously acquired mosquito-borne disease infection in humans in New Zealand in which a pathogen was transmitted by a female mosquito from a local vector population (Weinstein et al., 1995; Derraik and Maguire 2005). However, an eventual epidemic in New Zealand has been predicted in the long term (Weinstein et al., 1997; Derraik and Slaney 2007). When a mosquito-borne disease outbreak does occur in New

Zealand, it is likely to have severe economic and public health implications. Because New Zealand lacks a historical precedent, the human population is non-immune and thus susceptible to a "virgin soil" outbreak (Weinstein et al., 1997; Hearnden et al., 1999). Moreover, the general public has little knowledge of mosquito-borne diseases or anti-mosquito measures (i.e., preventing larval breeding habitat, minimizing exposure during common biting times) to minimize the risk of infection (Weinstein et al., 1994). These problems are exacerbated by the facts that in New Zealand few health professionals have experience in identifying mosquito-borne diseases, especially arboviral infections, and there is the lack of rapid and widespread diagnostic services (Weinstein 1995; Derraik and Calisher 2004). All of these factors combined are likely to result in the rapid spread and delayed recognition of mosquito-borne disease (Weinstein et al., 1997; Hearnden et al., 1999; Derraik and Calisher 2004).

While the economic cost of any disease outbreak is difficult to calculate, in 1994 New Zealand's Ministry of Health estimated the cost of a potential dengue outbreak in Auckland at US$180 million (Weinstein 1994), and in 2006, it was estimated that the initial cost of a virgin-soil outbreak of Ross River virus would be approximately US$89 million (MoH 2006). The establishment in New Zealand of *A. camptorhycus*, which could initiate this Ross River virus outbreak, is the reason for an extensive ongoing eradication program costing over US$38 million. Given *A. camptorhycus* is not a container breeder, the cost of controlling a resident species or eradicating a resident invader that is adapted to breeding in artificial containers would be substantially higher. Eradication of a container-breeding mosquito would be especially difficult in an urban setting where containers are widely dispersed, extremely various, and often cryptic in nature.

Urbanization increases the risk of mosquito-borne disease in New Zealand via three key mechanisms: (i) providing good entry points for the invasion and establishment of novel diseases and exotic mosquitoes, (ii) encroachment on remnant native forest areas that appear to provide habitat for exotic disease vectors, and (iii) providing favorable habitat for the proliferation and spread of native and exotic mosquitoes and pathogens. All of these mechanisms were explained earlier, and in the following sections we will discuss them further within the context of risks to New Zealand.

Mosquito and Disease Introductions in an Era of Globalization and Climate Change

New Zealand's relative isolation has limited the establishment of many exotic insects and pathogens that are prevalent in other countries. However, increasing globalization is compromising this local refractivity. Since World War II, New Zealand has emerged as a leading nation in the South Pacific region and a notable participant in the global economy, with its largest cities firmly situated on international air and shipping routes. International travel and trade is increasing the number of infected travelers and exotic mosquitoes entering New Zealand. For instance, Kelly-Hope et al., (2002) estimated that more than 100 people per annum enter New Zealand from Queensland, Australia, with clinical or subclinical infections of Ross River virus. Since 1929, 171 mosquito interceptions have also been detected in New Zealand (Derraik 2004a). Despite more stringent biosecurity since the 1990s than at any time before, there has been a disproportionately high percentage (36%) of mosquito interceptions during this period (Derraik 2004a). Because of the recent centralization of New Zealand's infrastructure into fewer major urban centers, almost all international air travel and shipping enters a limited number of ports; the most prominent being in New Zealand's most populous city, Auckland (pop. ~1.3 million).

The risk of invasion and establishment of novel mosquitoes and pathogens is not solely dependant on large-scale travel and trade. The awareness of global climate change has drawn attention to the link between New Zealand's current and potential future climate and vector-borne diseases (Laird 1990; de Wet et al., 2001; de Wet et al., 2005). Currently, suitable climate conditions appear to exist for the invasion, establishment, and spread of *A. albopictus*, *Aedes japonicus*, *Aedes polynesienis*, *Aedes vigilax*, and *Culex annulirostris* (de Wet et al., 2005). As discussed in a previous section of this chapter, New Zealand is warm enough for outbreaks of arboviruses and mosquito-borne disease such as filariasis. However, the current climate is unlikely to allow the establishment of *Anopheles* species that can transmit malaria perhaps except for the temperate climate malaria vector, *Anopheles annulipes* (Boyd and Weinstein 1996). Of particular concern is that pathogen/parasite transmission would be greatest in the northern half of the North Island where the majority of the population live, including Auckland city, because this is the warmest part of country (Derraik and Slaney 2007). National climate studies have noted that New Zealand's temperature has gradually increased over the last century (Royal Society of New Zealand 1990; NIWA 2007b). Recent climate change forecasts by the Intergovernmental Panel on Climate Change (IPCC) have predicted an overall increase in New Zealand's average temperatures, with average rainfall decreasing in some regions and increasing in others, and more frequent extreme climatic events (NIWA 2007a). These changes would allow an additional suite of vectors to potentially invade New Zealand, including the principal worldwide dengue/DHF vector, *A. aegypti*, as well as extending the potential distribution of invading exotic species or of those that are already established (de Wet et al., 2005).

Encroachment of Urban Areas into New Zealand's Native Forest Remnants

Recent sampling in relatively pristine native habitats using CO_2 baited light traps has indicated a mosquito fauna comprised almost entirely of native species (Derraik et al., 2005b; Derraik et al., 2005a). The only exotic mosquito recorded was *A. notoscriptus*, but at very low numbers and only close to the forest edge. However, most of New Zealand's remaining lowland native forests are isolated fragments surrounded by agricultural land and, most notably, expanding urban landscapes. Past authors have suggested that *A. notoscriptus*, a natural container breeder in its native Australia (Lee et al., 1984) and important potential disease vector (see previous sections in this chapter), appears to be invading peri-urban native habitats which likely provide favorable shade and water conditions (Laird 1990; Laird 1995; Leisnham 2005). Surveys of natural containers in remnant forest patches in Auckland and Wellington cities found exclusively *A. notoscriptus* (Derraik 2004b; Derraik 2005), while a study that ovitrapped across neighboring patches of various land use in the greater Wellington region found that *A. notoscriptus* was found in both residential and native areas in relatively equal numbers as *C. pervigilans* (Leisnham 2005). Although numerous species of non-culicid invertebrates breed in natural containers within forest fragments (Derraik and Heath 2005; Leisnham 2005), few endemic mosquito species and predators have been recorded (Belkin

1968; Laird 1995; Leisnham et al., 2005). Endemic mosquitoes have probably always underutilized available habitats because of a species-poor fauna and a temperate climate, but at least one endemic mosquito species, *Maorigoeldia argyropus,* appears to have become recently endangered (Snell et al., 2005). This may be driven by the destruction and isolation of native forests due to urbanization and agricultural development (Derraik and Slaney 2007), although this has not yet been investigated. The elimination of natural community-level population regulators, such as inter-specific competition and predation, likely facilitates the proliferation of *A. notoscriptus* and may increase the potential for exotic container breeders to invade New Zealand, including known container breeders that have been intercepted at New Zealand's ports (e.g., *A. albopictus* and *A. japonicus*).

Remnant native areas in close proximity to residential areas may provide a substantial source of female mosquitoes to transmit and sustain disease transmission. Mark-release-capture studies have shown that it is common for adult females of many species to fly over 5 km in Australia (e.g., Russell 1986), depending on a number of factors such as the density of hosts, wind direction, and landscape (Muir and Kay 1998; Clements 2000b). There are no detailed studies on the flight distances of *A. notoscriptus* or any other New Zealand mosquitoes. However, it is reasonable to suggest that relatively densely populated urban areas may attract adult mosquitoes from New Zealand's remnant native forests, which are characterized as having few mammals or birds to provide blood meals (Diamond 1990). The establishment of *A. notoscriptus* and other mosquito species in native forest patches is likely to make surveillance and control operations more difficult when an outbreak of vector-borne human disease occurs. Public interventions that aim to limit larval sites will be less effective at controlling the regional population size since the most important sites will not be targeted. Organized surveillance operations are likely to require increased resources since locating key larval habitats, such as water-filled natural containers, will be more difficult as they are often elevated above ground and cryptic. Moreover, native forest reserves in close proximity to urban areas are often subject to the illegal dumping of anthropogenic material (Bagnall 1979; Stenhouse 2004). Thus, disused artificial containers may supplement natural containers (e.g., treeholes) as excellent habitats for capable container-utilizing species like *A. notoscriptus*. Even when larval habitats are located in native forests the control of the mosquitoes within them using insecticides is likely to be prohibitive in many situations.

Favorable Habitats in New Zealand Cities

In the context of New Zealand's mosquito-borne disease risk, the most important environmental change caused by urbanization is the proliferation of favorable breeding habitats. New Zealand's most extensive mosquito larval survey was conducted in 1993–94 (Laird 1995). It found that 85% of all utilized habitats were artificial containers in modified (mainly urban) environments (Laird 1995). While this study gave priority to sampling such habitats, an earlier (and previously largest) study in the northern parts of New Zealand in 1988–89 also showed that artificial containers were the most common (65%) and utilized (65%) larval habitat (Laird 1990). Like most urban areas overseas, New Zealand's cities contain a variety of container types, including plant pots, bird baths, storm water drains, and any kind of disused container that could fill with rain water (Laird 1990; Laird 1995; Leisnham et al., 2005). Laird (1995) recorded 37 varieties of container habitats used in New

Zealand. Particularly noteworthy are used tire casings. Laird (1995) ascribed their likely role, in combination with brief warm periods, for the range expansion of both *A. notoscriptus* and *C. quinquefasiatus* southward throughout the North Island. Both species have now reached Christchurch, half way down the South Island.

Aedes notoscriptus and *C. quinquesfasiatus* were the second and third most abundant species collected in the 1993-94 survey, respectively (Laird 1995). Although *C. pervigilans* was still New Zealand's most common and widespread species, less than 1% of the total larvae recorded were other endemic species. These results revealed that artificial containers are particularly underutilized by endemic species and thus present unfilled ecological niche space for the expansion of exotic species currently in New Zealand and for those that threaten to invade (Laird 1990).

Citing recent studies that surveyed resident containers and used experimental oviposition traps (Derraik 2004b; Derraik et al., 2005a), Derriak and Slaney (2007) suggested that *A. notoscriptus* and *C. quinquesfastiatus* may now be replacing *C. pervigilans* as the most dominant mosquito in urban and peri-urban areas in the Auckland region. Ovitrapping studies suggest that *C. pervigilans* is still the dominant species in urban areas in the Wellington region (Lester and Pike 2003; Leisnham et al., 2005; Leisnham et al., 2006) and probably experience more favorable larval conditions in urban areas than in native forest, with higher temperatures (Leisnham et al., 2006) and nutrient levels (Leisnham et al., 2004; Leisnham et al., 2005), and fewer predators (Leisnham et al., 2004).

What We Need to Learn about New Zealand's Urban Mosquitoes

Despite no historical precedent, urbanization is increasing the risk of mosquito-borne human disease in New Zealand by favouring the invasion, establishment, and spread of exotic mosquitoes and pathogens. A number of authors have commented that New Zealand does not appear to have any particular characteristics that make it fundamentally refractory to the local transmission of arboviruses (e.g., Weinstein et al., 1997), and the absence of any records of a mosquito-borne human arboviral infection in New Zealand seems to be entirely fortuitous (Weinstein et al., 1997; Derraik and Calisher 2004). Exotic species already established in New Zealand present an immediate public health risk should a disease be introduced. Although it is of course impossible to directly study the effects of urbanization on exotic mosquitoes before they invade, species currently in the country provide models to help forecast the impacts and responses of future invaders that are superior vectors. Therefore, information on the ecology of New Zealand's mosquitoes is now vital for efficient surveillance and control measures given the risk they pose.

Important advances have been made in the last 10 years at identifying the habitat preferences of New Zealand's mosquitoes. This research has clearly shown that urban areas provide favorable habitat for at least two of New Zealand's exotic species, and will likely do so for more dangerous vectors that threaten to invade. However, we still know little of two important components of the urbanization–mosquito relationship in New Zealand. The first component is the quantitative description of the changes in population number and form of population growth or decline that coincide with urbanization. Longitudinal data on mosquito densities before, during and after urbanization have not been collected in New Zealand and would provide direct data on the effects of urbanization. Descriptive data on vector population dynamics is important because it provides the data that can answer important questions about control efforts (e.g., Are control efforts justified at this time? What is the expected population of a mosquito species some time in the future and is some intervention, such as controlling breeding habitat, desirable to alter that expectation?). The second component deals with causal mechanisms, and is important because it can provide a general framework for strategies to control mosquitoes, including future invaders (Juliano 2007). Knowledge of causal processes affecting population dynamics also may improve the forecasts of population trends (e.g., Does the elimination of a certain predator lead to increases in one species and not another? Why is this?).

Of particular importance is answering the questions: (i) Why are *A. notoscriptus* and *C. quinquefasiastus* appearing to be becoming the dominant urban species in Auckland, (ii) Is this dominance occurring in other parts of New Zealand (e.g., Wellington) and, if not, why? Descriptive data from New Zealand populations and experimental data from overseas both identify climate as a likely important variable affecting expansion of *A. notoscriptus* and *C. quinquefasiatus* (de Wet et al., 2005) as well as the establishment of other potential invaders. However, the impacts and responses to climatic variables of New Zealand's exotic mosquitoes are based on knowledge from overseas populations. Intra-specific populations of mosquitoes from geographically discrete locations often exhibit phenotypic differences because of differential local selection pressures (e.g., Lounibos et al., 1993; Frankino and Juliano 1999; Armbruster et al., 2001; Armbruster and Conn 2006), and it is likely that New Zealand populations of *A. notoscriptus* and *C. quinquefasiatus* respond differently to climate and other environmental variables than populations overseas. Furthermore, while surveys have identified mosquito species that co-occur in larval habitats (e.g., Laird 1995; Derraik 2005) and their associated invertebrate communities (Derraik and Heath 2005; Leisnham 2005), we have no data on the inter-specific interactions (e.g., competition, predation) that may limit population densities. Investigations of causal mechanisms that drive population densities require experimental studies in field and control laboratory settings. These experiments represent the next step for future research in answering important questions on the roles of climate and inter-specific interactions and other abiotic and biotic factors in driving vector densities in New Zealand.

Conclusion

Descriptive data on the distribution and abundance of vectors and their pathogens is the focus of much disease epidemiology, and plays an important role in vector control and public health interventions. The mechanisms by which urbanization increases the risk of mosquito-borne diseases is only beginning to be understood. A better knowledge of the responses and impacts of mosquitoes to a range of physical, chemical, and biological gradients would allow more accurate predictions of the change in risk exposure to mosquitoes that would be expected after urbanization. Because environmental factors are likely to be highly interlinked, further study on the response of mosquito populations to them would ideally involve experimentally manipulated studies in both field and laboratory settings to test their relative effects. Ecological data then needs to be incorporated into models that include other important variables, such as climatic parameters, topographic details, and the proximity of seaports and airports, as well as information on human population density and socio-economic status, to identify scenarios that lead to particularly high disease risk in urban areas.

However, identifying scenarios that lead to high disease risk in urban areas is only addressing one part of the public-health problem. Preparing biosecurity and health workers to identify and treat mosquito-borne diseases and respond to outbreaks is essential. Choosing cost-effective interventions that lower disease risk and are logistically implementable and sustainable is also necessary for mitigating mosquito-borne diseases. This involves an understanding of the complex trade-offs involved when considering human health in regards to mosquito-borne disease. While some diseases proliferate as a result of poor urban planning (e.g., increased dengue risk with poor sanitation), the risk of other diseases may increase with 'good' urban planning. For instance, the protection or restoration of native systems within peri-urban landscapes fulfill other important ecological and human health purposes (e.g., wildlife habitat, landscape enhancement, water storage and/or flood mitigation) (USEPA 2000; Horwitz et al., 2001), yet may also facilitate the proliferation of mosquito-borne diseases. Understanding the ecology of specific mosquito species will help balance the multi-purpose goals of native forests with their potential in providing mosquito larval habitat.

The urbanization-mosquito relationship highlights the two-way link between human well-being and ecosystem health and supports calls from authors in the last 15 years for the further convergence of ecology and health paradigms (e.g., Hales et al., 1997; Weinstein 1997; Parkes et al., 2003; Parkes and Weinstein 2004; Millennium Ecosystem Assessment 2005; Pongsiri and Roman 2007). Traditional disease models can cope with the direct effect of an urban-mediated exposure for a specific disease outcome; but the urbanization-mosquito relationship demands a multidisciplinary approach to understand the indirect relationships between vectors, pathogens, humans, and ecological disturbance. Thus, while cities face the increasing public health threat from degraded ecosystems and mosquito-borne disease as they enter the 21st century, they also potentially provide the nurturing ground for the necessary information sharing, creativity, and innovation needed to preserve human well-being.

Acknowledgements

I thank P.J. Lester and B. Kesavaraju for comments on this review, and S.A. Juliano for supporting the development and time required for this review.

References

Abe, M., McCall, P. J., Lenhart, A., Villegas, E., and Kroeger, A. (2005). The Beun Pastor Cemetery in Trujillo, Venezuela: Measuring dengue vector output from a public area. *Tropical Medicine and International Health,* **10**, 597-603.

Afrane, Y. A., Klinkenberg, E., Drechsel, P., Owusu-Daaku, K., Garms, R., and Kruppa, T. (2004). Does irrigated urban agriculture influence the transmission of malaria in the city of Kumasi, Ghana? *Acta Tropica,* **89**, 125-134.

Anderson, R. M., and May, R. M. (1991). *Infectious Diseases of Humans. Dynamics and Control.* Oxford, UK: Oxford University Press.

Annan, K. A. (2000). *Millennium Report of the Secretary-General of the United Nations.* New York, NY: United Nations.

Antonio-Nkondjio, C., Simard, F., Awono-Ambene, P., Ngassam, P., Toto, J. C., Tchuinkam, T., and Fontenille, D. (2005). Malaria vectors and urbanization in the equatorial forest region of south Cameroon. *Transactions of the Royal Society of Tropical Medicine and Hygiene,* **99**, 347-354.

Armbruster, P., Bradshaw, W. E., Ruegg, K., and Holzapfel, C. M. (2001). Geographic variation and the evolution of reproductive allocation in the pitcher-plant mosquito, *Wyeomyia smithii. Evolution,* **55**, 439-444.

Armbruster, P., and Conn, J. E. (2006). Geographic variation of larval growth in North American *Aedes albopictus* (Diptera: Culicidae). *Annals of the Entomological Society of America,* **99**, 1234-1243.

Aron, J. L., and Patz, J. A., (Eds.). (2001). *Ecosystem Health and Public Health.* Baltimore, MD: Johns Hopkins University Press.

Bagnall, R. G. (1979). Study of human impact on an urban forest remnant - Redwood Bush, Tawa, near Wellington, New Zealand. *New Zealand Journal of Botany,* **17**, 117-126.

Barrett, R., Kuzawa, C. W., McDade, T., and Armelagos, G. J. (1998). Emerging and re-emerging infectious diseases: The third epidemiologic transition. *Annual Review of Anthropology,* **27**, 247-271.

Begon, M., Harper, J. L., and Townsend, C. R. (1990). *Ecology: Individuals, Populations and Communities.* (2nd edition). Cambridge, UK: Blackwell Scientific Publications.

Belkin, J. N. (1968). Mosquito studies (Diptera, Culicidae) VII. The Culicidae of New Zealand. *Contributions of the American Entomological Institute,* **3**, 1-182.

Boyd, A. M., and Weinstein, P. (1996). *Anopheles annulipes* Walker s.l. (Diptera: Culicidae), an under-rated temperate climate malaria vector? *New Zealand Entomologist,* **19**, 35-41.

Bradley, C. A., and Altizer, S. (2006). Urbanization and the ecology of wildlife diseases. *Trends in Ecology and Evolution,* **22**, 95-102.

Case, T. J. (1990). Invasion resistance arises in strongly interacting species-rich model competition communities. *Proceedings of the National Academy of Sciences of the United States of America, 87*, 9610-9614.

CDC (2007). *Final 2006 West Nile Virus Activity in the United States.* Atlanta, USA. Available: *http://www.cdc.gov/ncidod/dvbid/westnile/Mapsactivity/survandcontrol06Maps.htm* [accessed 31 May 2007], Centers for Disease Control and Prevention.

Chivian, E. (2001). Environment and health: 7. Species loss and ecosystem disruption the implications for human health. *Canadian Medical Association Journal,* **164**, 66-69.

Clements, A. N. (2000a). *The Biology of Mosquitoes: Development, Nutrition and Reproduction.* Wallingford, UK: CABI Publishing.

Clements, A. N. (2000b). *The Biology of Mosquitoes: Sensory Reception and Behaviour.* Wallingford, UK: CABI Publishing.

Coluzzi, M. (1994). Malaria and the Afrotropical ecosystems: Impact of man-made environmental changes. *Parassitologia,* **36**, 223-227.

Cook, A., Weinstein, P., and Woodward, A. (2002). The impact of exotic insects in New Zealand. In Pimentel, D. (Ed.), *Biological Invasions: Economic and Environmental Costs of Alien Plant, Animal, and Microbe Species.* (pp. 217-239). Boca Raton, USA: CRC Press.

Costanzo, K. S., Kesavaraju, B., and Juliano, S. A. (2005). Condition specific competition in container mosquitoes: The role of non-competing life-history stages. *Ecology,* **86**, 3289-3295.

Cowan, P. E. (1990). Brushtail possum. In King, C. M. (Ed.), *The Handbook of New Zealand Mammals.* (pp. 68-98). Auckland, NZ: Oxford University Press.

Daszak, P., Cunningham, A. A., and Hyatt, A. D. (2000). Emerging infectious diseases of wildlife-threats to biodiversity and human health. *Science,* **287**, 443-449.

Daszak, P., Cunningham, A. A., and Hyatt, A. D. (2001). Anthropogenic environmental change and the emergence of infectious diseases in wildlife. *Acta Tropica,* **78**, 103-116.

Daszak, P., Tabor, G. M., Kilpatrick, A. M., and Epstein, J. (2004). Conservation medicine and a new agenda for emerging diseases. *Annals of the New York Academy of Sciences, 1026*, 1-11.

Day, J. F. (2001). Predicting St. Louis encephalitis epidemics: Lessons from recent and not so recent outbreaks. *Annual Review of Entomology,* **46**, 111-138.

De, S. K., and Chandra, G. (1994). Studies on the filariasis vector - *Culex quinquefasciatus* at Kanchrapara, West-Bengal (India). *Indian Journal of Medical Research,* **99**, 255-258.

de Wet, N., Slaney, D. P., Ye, W., Warrick, R., and Hales, S. Hotspots: Exotic mosquito risk profiles for New Zealand [online]. (2005). Available from: *http://waikato.ac.nz/igic/hotspots* [accessed 26 May 2007].

de Wet, N., Ye, W., Hales, S., Warrick, R., Woodward, A., and Weinstein, P. (2001). Use of a computer model to identify potential hotspots for dengue fever in New Zealand. *New Zealand Medical Journal,* **114**, 420-422.

Depinay, J.-M., Mbogo, C., Killeen, G., Knols, B., Beier, J., Carlson, J., Dushoff, J., Billingsley, P., Mwambi, H., Githure, J., Toure, A., and Ellis McKenzie, F. (2004). A simulation model of African *Anopheles* ecology and population dynamics for the analysis of malaria transmission. *Malaria Journal,* **3,** 29.

Derraik, J., Snell, A., and Slaney, D. (2005a). An investigation into the time of activity of adult mosquitoes (Diptera: Culicidae) seeking host-cues in West Auckland. *New Zealand Entomologist,* **28**, 85-90.

Derraik, J., Snell, A., and Slaney, D. (2005b). Vertical distribution of mosquitoes in native forest in Auckland, New Zealand. *Journal of Vector Ecology,* **30**, 334-336.

Derraik, J. G. B. (2004a). Exotic mosquitoes in New Zealand: A review of species intercepted, their pathways and ports of entry. *Australian and New Zealand Journal of Public Health,* **28**, 433-444.

Derraik, J. G. B. (2004b). A survey of the mosquito (Diptera: Culicidae) fauna of the Auckland Zoological Park. *New Zealand Entomologist,* **27**, 51-55.

Derraik, J. G. B. (2005). Mosquitoes breeding in phytotelmata in native forests in the Wellington region, New Zealand. *New Zealand Journal of Ecology,* **29**, 185-191.

Derraik, J. G. B., and Calisher, C. H. (2004). Is New Zealand prepared to deal with arboviral diseases? *Australian and New Zealand Journal of Public Health,* **28**, 27-30.

Derraik, J. G. B., and Heath, A. C. G. (2005). Immature Diptera (excluding Culicidae) inhabiting phytotelmata in the Auckland and Wellington regions. *New Zealand Journal of Marine and Freshwater Research,* **39**, 981-987.

Derraik, J. G. B., and Maguire, T. (2005). Mosquito-borne diseases in New Zealand: Has there ever been an indigenously acquired infection? *New Zealand Medical Journal,* **118**, 1670.

Derraik, J. G. B., and Slaney, D. P. (2007). Anthropogenic environmental change, mosquito-borne diseases and human health in New Zealand. *EcoHealth,* **4,** 72-81.

Derraik, J. G. B., and Snell, A. E. (2004). Notes on daytime biting catches of mosquitoes (Diptera: Culicidae) in native forest sites in the Auckland region. *The Weta,* **28**, 14-19.

Diamond, J. M. (1990). New Zealand as an archipelago: An international perspective. In Towns, D. R., Daugherty, C. H. and Atkinson, I. A. E. (Eds.), *Ecological Restoration of New Zealand Islands*. (pp. 3-8). Wellington: Department of Conservation.

DoC, and MfE (2000). *New Zealand Biodiversity Strategy.* Wellington, New Zealand, Department of Conservation and Ministry for the Environment.

Doggett, S. L., and Russell, R. C. (1997). *Aedes notoscriptus* can transmit inland and coastal isolates of Ross River and Barmah Forest viruses from New South Wales. *Arbovirus Research in Australia,* **7**, 79-81.

ESR (2007). *Notifiable and Other Diseases in New Zealand, Annual Report.* Wellington, NZ, ESR.

Ezenwa, V. O., Godsey, M. S., King, R. J., and Guptill, S. C. (2006). Avian diversity and West Nile virus: Testing associations between biodiversity and infectious disease risk. *Proceedings of the Royal Society of Biological Sciences,* **273**, 109-117.

Faeth, S. H., Warren, P. S., Shochay, E., and Marussich, W. A. (2005). Trophic dynamics in urban communities. *Bioscience,* **55**, 399-407.

Fanning, I. D., Crisp, G., and Mottram, P. (1997). Bromeliads as breeding sites for mosquitoes in south-east Queensland. *Arbovirus Research in Australia*, **7,** 82-83.

Folke, C., Jansson, A., Larsson, J., and Costanza, R. (1997). Ecosystem appropriation by cities. *Ambio,* **26**, 167-172.

Forget, G., and Lebel, J. (2001). An ecosystem approach to human health. *International Journal of Occupational and Environmental Health*, **7**, S3-S36.

Frankie, G. W., and Ehler, L. E. (1978). Ecology of insects in urban environments. *Annual Review of Entomology,* **23**, 367-387.

Frankino, W. A., and Juliano, S. A. (1999). Costs of reproduction and geographic variation in the reproductive tatics of the mosquito *Aedes triseriatus. Oecologia,* **120**, 59-68.

Gerhardt, R. R., Gottfried, K. L., Apperson, C. S., Davis, B. S., Erwin, P. C., Smith, A. B., Panella, N. A., Powell, E. E., and Nasci, R. S. (2001). The first isolation of La Crosse virus from naturally occurring infected *Aedes albopictus. Emerging Infectious Diseases,* **7,** 807-811.

Gilglioli, G. (1963). Ecological change as a factor in renewed malaria transmission on the Damerara River estuary, British Guinea, in the fifteenth year of *A. darlingi* and malaria eradication. *Bulletin of the World Health Organization,* **29**, 131-145.

Gratz, N. G. (1997). Dengue and dengue hemorrhagic fever: A spreading pandemic. *Public Health,* **14**, 20-29.

Gratz, N. G. (1999). Emerging and resurging vector-borne diseases. *Annual Review of Entomology,* **44**, 51-75.

Haines, A., McMichael, A. J., and Epstein, P. R. (2000). Environment and health: 2. Global climate change and health. *Canadian Medical Association Journal,* **163**, 729-734.

Hales, S., Weinstein, P., and Woodward, A. (1997). Public health impacts of global climate change. *Reviews on Environmental Health,* **12**, 191-200.

Halstead, S. B. (1993). Global epidemiology of dengue: Health systems in disarray. *Tropical Medicine,* **35**, 137-146.

Hearnden, M., Skelly, C., Dowler, H., and Weinstein, P. (1999). Improving the surveillance of mosquitoes with disease-vector potential in New Zealand. *New Zealand Public Health Report,* **6**, 25-30.

Holder, P., Brown, G., and Bullians, M. (1999). The mosquitoes of New Zealand and their animal disease significance. *Surveillance,* **26**, 12-15.

Horwitz, P., Lindsay, M., and O'Conner, M. (2001). Biodiversity, endemism, sense of place, and public health: Inter-relationships for Australian inland aquatic systems. *Ecosystem Health,* **7,** 253-265.

Ibañez-Berñal, S., B., Briseño, J. P., Mutebi, E. A., and Rodriguez, G. (1997). First record in America of *Aedes albopictus* naturally infected with dengue virus during the 1995 outbreak at Reynosa, Mexico. *Medical and Veterinary Entomology,* **11**, 305-309.

Juliano, S. A. (1998). Species introduction and replacement among mosquitoes: Interspecific resource competition or apparent competition. *Ecology,* **79**, 255-268.

Juliano, S. A. (2007). Population dynamics. In T. G. Floore (Ed.), *Biorational Control of Mosquitoes*. Supplement to the *Journal of the Mosquito Control Association,* **23**, 265-275. American Mosquito Control Association, Bulletin No. 7.

Juliano, S. A., and Lounibos, L. P. (2005). Ecology of invasive mosquitoes: Effects on resident species and on human health. *Ecology Letters*, **8,** 558-574.

Juliano, S. A., Lounibos, L. P., and O'Meara, G. F. (2004). A field test for competitive effects of *Aedes albopictus* on *A. aegypti* in South Florida: Differences between sites of coexistence and exclusion. *Oecologia,* **194**, 583-593.

Juliano, S. A., O'Meara, G. F., Morrill, J. R., and Cutwa, M. M. (2002). Desiccation and thermal tolerance of eggs and the coexistence of competing mosquitoes. *Oecologia,* **130**, 458-469.

Kay, B. H., Ryan, D., Russel, B. M., Holder, J. S., Lyons, J., and Foley, P. N. (2000). The importance of subterranean mosquito habitat to arbovirus vector control strategies in North Queensland, Australia. *Journal of Medical Entomology,* **37**, 846-853.

Kelly-Hope, L. A., Kay, B. H., Purdie, D. M., and Williams, G. M. (2002). The risk of Ross River and Barmah Forest virus disease in Queensland: Implications for New Zealand. *Australian and New Zealand Journal of Public Health,* **26**, 69-77.

Kelly-Hope, L. A., Rolain, J. M., and Raoult, D. (2005). Prevalence of human pathogens in cat and dog fleas in New Zealand. *New Zealand Medical Journal,* **118**, U1754.

Kilpatrick, A. M., Kramer, L. D., Jones, M. J., Marra, P. P., and Daszak, P. (2006). West Nile virus epidemics in North America are driven by shifts in mosquito feeding behavior. *Public Library of Science,* **e82**.

Koenraadt, C. J., and Takken, W. (2003). Cannibalism and predation among larvae of the *Anopheles gambiae* complex. *Medical and Veterinary Entomology,* **17**, 61-66.

Lacey, L. A., and Lacey, C. M. (1990). The medical importance of riceland mosquitoes and their control using alternatives to chemical pesticides. *Journal of American Mosquito Control Association,* **6**, 2.

Laird, M. (1990). New Zealand's Northern Mosquito Survey, 1988-89. *Journal of the American Mosquito Control Association,* **6**, 287-299.

Laird, M. (1995). Background and findings of the 1993-94 New Zealand Mosquito Survey. *New Zealand Entomologist,* **18**, 77-91.

Laird, M., Calder, L., Thornton, R. C., Syme, R., Holder, P. W., and Mogi, M. (1994). Japanese *Aedes albopictus* among four mosquito species reaching New Zealand in used tires. *Journal of the American Mosquito Control Association,* **10**, 14-23.

Laird, M., and Easton, J. M. (1994). *Aedes notoscriptus* (Diptera: Culicidae) in Wellington Province. *New Zealand Entomologist,* **17**, 14-17.

Lee, D. J., Hicks, M. M., Griffiths, M., Russell, R. C., and Marks, E. H. (1984). *The Culicidae of the Australasian Region.* Canberra, Australia: Australian Government Publishing Service.

Leisnham, P. T. (2005). *Anthropogenic land use change and mosquitoes of public health importance.* Dunedin, NZ, University of Otago. PhD thesis.

Leisnham, P. T., Lester, P. J., Slaney, D. P., and Weinstein, P. (2004). Anthropogenic landscape change and vectors in New Zealand: The effects of shade and nutrient levels on mosquito productivity. *EcoHealth,* **1**, 306-316.

Leisnham, P. T., Lester, P. J., Slaney, D. P., and Weinstein, P. (2006). Relationships between mosquito densities in artificial container habitats, land use and temperature in the Kapiti-Horowhenua region, New Zealand. *New Zealand Marine and Freshwater Research,* **40**, 285-298.

Leisnham, P. T., Slaney, D. P., Lester, P. J., and Weinstein, P. (2005). Increased larval mosquito densities from modified landuses in the Kapiti region, New Zealand: Vegetation, water quality and predators as associated factors. *EcoHealth,* **2,** 1-10.

Lester, P. J., and Pike, A. J. (2003). Container diameter and water depth influence population dynamics of the mosquito *Culex pervigilans* (Diptera: Culicidae) and its associated predators in New Zealand. *Journal of Vector Ecology,* **28**, 267-274.

Levine, J. M., and D'Antonio, C. M. (1999). Elton revisited: A review of evidence linking diversity and invasibility. *Oikos,* **87**, 15-26.

Lindblade, K. A., Walker, E. D., Onapa, A. W., Katungu, J., and Wilson, M. L. (2000). Land use change alters malaria transmission parameters by modifying temperature in a highland area of Uganda. *Tropical Medicine and International Health,* **5,** 263-274.

Lindsay, M. D., Oliveira, N., Jasinska, E., Johansen, C., Harrington, S., Wright, A. E., and Smith, D. (1996). An outbreak of Ross River Virus disease in southwestern Australia. *Emerging Infectious Diseases,* **2,** 117-120.

Lindsay, S. W., and Birley, M. H. (1996). Climate change and malaria transmission. *Annals of Tropical Medicine and Parasitology,* **90**, 573-586.

Livernash, R., and Rodenburg, E. (1998). *Population Bulletin* Vol 53 No 1, *Population Change, Resources, and the Environment.* Population Bulletin. Washington DC, USA., Population Reference Bureau.

Ljumba, J. N., and Lindsay, S. W. (2001). Impact of irrigation on malaria in Africa: Paddies paradox. *Journal of Medical and Veterinary Entomology,* **15**, 1-11.

Ljumba, J. N., Mosha, F. W., and Lindsay, S. W. (2002). Malaria transmission risk variations derived from different agricultural practices in an irrigated area of northern Tanzania. *Medical and Veterinary Entomology,* **16**, 28-38.

LoGiudice, K., Ostfeld, R. S., Schmidt, K. A., and Keesing, F. (2003). The ecology of infectious disease: Effects of host diversity and community composition on Lyme disease risk. *Proceedings of National Academy of Sciences of the USA,* **100**, 567-571.

Lounibos, L. P. (1985). Interactions influencing production of treehole mosquitoes in south Florida. In Lounibos, L. P., Rey, L. R. and Frank, J. H. (Eds.), *Ecology of Mosquitoes: Proceedings of a Workshop.* (pp. 65-77). Vero Beach, FL: Florida Medical Entomology Laboratory.

Lounibos, L. P. (2002). Invasions by insect vectors of human disease. *Annual Review of Entomology,* **47**, 233-266.

Lounibos, L. P., Escher, R. L., and Lourenço-de-Oliveira, R. (1993). Asymmetric evolution of photoperiodic diapause in temperate and tropical invasive populations of *Aedes albopictus* (Diptera: Culicidae). *Annals of the Entomological Society of America,* **96**, 512-518.

Lundstrom, J., Vene, J., Espmark, A., Engvall, A., and Miklasson, B. (1991). Geographical and temporal distribution of Ockelbo disease in Sweden. *Epidemiological Infections,* **106**, 567-574.

Mackenzie, J. S., Gubler, D. J., and Petersen, L. R. (2004). Emerging flaviviruses: the spread and resurgence of Japanese encephalitis, West Nile and dengue viruses. *Nature Medicine,* **10**, S98-S109.

Maguire, T., Miles, J. A. R., and Casals, J. (1967). Whataroa virus, a group A arbovirus isolated in South Westland, New Zealand. *American Journal of Tropical Medicine and Hygiene,* **16**, 371-373.

Massey, B., Gleeson, D. M., Slaney, D., and Tompkins, D. M. P. (2007). CR detection of Plasmodium and blood-meal identification in a native New Zealand mosquito. *Journal of Vector Ecology,* **32**, 154-156.

Matthys, B., N'Goran, E. K., Kone, M., Koudou, B. G., Vounatsou, P., Cisse, G., Tschannen, A. B., Tanner, M., and Utzinger, J. (2006). Urban agricultural land use and characterization of mosquito larval habitats in a medium-sized town of Cote d'Ivoire. *Journal of Vector Ecology,* **31**, 319-333.

May, R. M., and Anderson, R. M. (1979). Population biology of infectious diseases: Part II. *Nature,* **280**, 455-461.

McIntyre, N. E. (2000). Ecology of urban arthropods: A review and a call to action. *Annals of the Entomological Society of America,* **93**, 825-835.

McLean, D. M. (1953). Transmission of Murray Valley encephalitis virus by mosquitoes. *Australian Journal of Experimental Biology and Medical Science,* **31**, 481-490.

McMichael, A. J., and Haines, A. (1997). Global climate change: The potential effects on health. *British Medical Journal,* **315**, 805-809.

Merritt, R. W., Dadd, R. H., and Walker, E. D. (1992). Feeding-behavior, natural food, and nutritional relationships of larval mosquitoes. *Annual Review of Entomology,* **37**, 349-376.

Meyerson, F. A. B., Merino, L., and Durand, J. (2007). Migration and environment in the context of globalization. *Frontiers in Ecology and the Environment,* **5,** 182-189.

MfE (1997). *The State of New Zealand's Environment.* Wellington, NZ, Ministry for the Environment.

MfE (2006). *Gentle Footprints Boots 'n' All.* Wellington, NZ. Available: *http://www.mfe.govt.nz/publications/ser/gentle-footprints-may06/index.html* [accessed 15 June 2007]. Ministry for the Environment.

Miles, J. A. R. (1973). The ecology of the Whataroa virus, an alphavirus, in South Westland, New Zealand. *Journal of Hygiene,* **73**, 701-713.

Millennium Ecosystem Assessment (2005). *Ecosystems and Human Well-Being: Synthesis.* Washington, DC, Island Press.

MoH (2006). *A Health Risk Assessment Relating to the Establishment of the Exotic Mosquito Ochlerotatus camptorhynchus in the Southern Kaipara, Coromandel and Wairau.* Wellington, NZ, Ministry of Health.

Muir, L. E., and Kay, B. H. (1998). *Aedes aegypti* survival and dispersal estimated by mark-release-recapture in northern Australia. *American Journal of Tropical Medicine,* **58**, 277-282.

NIWA (2007a). *Climate Change. IPCC Fourth Assessment Report.* Wellington, NZ. Available: *http://www.niwa.co.nz/media/ipcc_4.pdf* [accessed 15 June 2007], National Institute of Water and Atmospheric Research.

NIWA (2007b). *Mean Annual Temperatures over New Zealand,* 1855-2005. Wellington, NZ. Available: *http://www.niwascience.co.nz/ncc/clivar/pastclimate* [accessed 15 June 2007], National Institute of Water and Atmospheric Research.

Norris, D. E. (2004). Mosquito-borne diseases as a consequence of land use change. *EcoHealth,* **1**, 19-24.

O'Meara, G. F. (1995). Spread of *Aedes albopictus* and decline of *Ae. aegypti* (Diptera: Culicidae) in Florida. *Journal of Medical Entomology,* **32**, 554-562.

Old, J. M., and Deane, E. M. (2005). Antibodies to the Ross River virus in captive marsupials in urban areas of eastern New South Wales, Australia. *Journal of Wildlife Diseases,* **41**, 611-614.

Ostfeld, R. S., and Keesing, F. (2000). The function of biodiversity in the ecology of vector-borne zoonotic disease. *Canadian Journal of Zoology,* **78**, 2061-2078.

Parkes, M., Panelli, R., and Weinstein, P. (2003). Converging paradigms for environmental health theory and practice. *Environmental Health Perspectives,* **111**, 669-675.

Parkes, M., and Weinstein, P. (2004). An ecosystem approach to environmental health. In Cromar, N., Cameron, S. and Fallowfield, H. (Eds.), *Environmental Health in Australia and New Zealand.* (pp. 45-65). Melbourne, Australia: Oxford University Press.

Parliamentary Commissioner for the Environment (2000). *Caught in the Headlights: New Zealand's Reflections on Possumns, Control Options and Genetic Engineering.* Wellington, NZ, Parliamentary Commissioner for the Environment.

Patz, J. A., Daszak, P., Tabor, G. M., Aguirre, A. A., Pearl, M., Epstein, J., Wolfe, N. D., Kilpatrick, A. M., Foufopoulos, J., Molyneux, D. H., Bradly, D. J., and Members of the Working Group on Land Use Change and Disease Emergence (2004). Unhealthy landscapes: Policy recommendations on land use change and infectious disease emergence. *Environmental Health Perspectives,* **112**, 1092-1098.

Patz, J. A., Graczyk, T. K., Geller, N., and Vittor, A. Y. (2000). Effects of environmental change on emerging parasitic diseases. *International Journal for Parasitology,* **30**, 1395-1405.

Pauchard, A., Aguayob, M., Peñaa, E., and Urrutiab, R. (2006). Multiple effects o urbanization on the biodiversity of developing countries: the case of a fast-growing metropolitan area (Concepcion, Chile). *Biological Conservation,* **127**, 272-281.

Petersen, L. R., and Marfin, A. A. (2005). Shifting epidemiology of flaviviridae. *Journal of Travel Medicine,* **12**, S3-S11.

Peterson, G., Allen, C. R., and Holling, C. S. (1998). Ecological resilience, biodiversity, and scale. *Ecosystems,* **1**, 6-18.

Pirages, D. C., Runci, P. J., and Sprinkle, R. H. (2001). Human populations in the shared environment. In Aron, J. L. and Patz, J. A. (Eds.), *Ecosystem Change and Public Health.* (pp. 165-187). Baltimore, USA: The Johns Hopkins University Press.

Polack, J. S. (1838). *New Zealand: Being a Narrative of Travels and Adventures during Residence in that Country Between the Years 1831 and 1837.* Volume 1. London, UK, Richard Bentley.

Pongsiri, M. J., and Roman, J. (2007). Examining the links between biodiversity and human health: an interdisciplinary research initiative at the U.S. environmental protection agency. *EcoHealth,* **4**, 82-85.

Reed, A. W. (1963). *An Illustrated Encyclopaedia of Maori Life.* Wellington, NZ: A.H. Reed and A.W. Reed.

Regis, L., Silvafilha, M. H. N., Deoliveira, C. M., Rios, E. M., Dasilva, S. B., and Furtado, A. F. (1995). Integrated control measures against *Culex quinquefasciatus*, the vector of filariasis in Recife. *Memorias Do Instituto Oswaldo Cruz,* **90**, 115-119.

Rosenberg, M. W., and Jaeger, J. (2000). Setting an agenda for research on health and the environment. *Global Change and Human Health,* **1**, 88-89.

Royal Society of New Zealand (1990). *New Zealand Climate Report* 1990. Bulletin No. 28. Wellington, NZ, The Royal Society of New Zealand.

Rubio-Palis, Y., and Zimmerman, R. H. (1997). Ecoregional classification of malaria vectors in the neotropics. *Journal of Medical Entomology,* **34**, 499-510.

Russell, B. M., McBride, W. J. H., Muller, H., and Kay, B. H. (2002). Epidemiological significance of subterranean *Aedes aegypti* (Diptera: Culicidae) breeding sites to dengue virus infection in Charters Towers, 1993. *Journal of Medical Entomology,* **39**, 143-145.

Russell, R. C. (1986). Dispersal of the arbovirus vector *Culex annulirostris* Skuse (Diptera: Culicidae) in the Murray Valley of Victoria Australia. *General and Applied Entomology,* **17**, 5-8.

Russell, R. C. (1995). Arboviruses and their vectors in Australia: an update on the ecology and epidemiology of some mosquito-borne arboviruses. *Review of Medical and Veterinary Entomology,* **83**, 141-158.

Russell, R. C. (2002). Ross River virus: Ecology and distribution. *Annual Review of Entomology,* **47**, 1-31.

Sala, O. E., Chapin, F. S., Armesto, J. J., Berlow, E., Bloomfield, J., Dirzo, R., Huber-Sanwald, E., Huenneke, L. F., Jackson, R. B., Kinzig, A., Leemans, R., Lodge, D. M., Mooney, H. A., Oesterheld, M., Poff, N. L., Sykes, M. T., Walker, B. H., Walker, M., and Wall, D. H. (2000). Biodiversity: Global biodiversity scenarios for the year 2100. *Science,* **287**, 1770-1774.

Sattler, M. A., Mtasiwa, D., Kiama, M., Premji, Z., Tanner, M., Killeen, G. F., and Lengeler, C. (2005). Habitat characterization and spatial distribution of *Anopheles* sp. mosquito larvae in Dar es Salaam (Tanzania) during an extended dry period. *Malaria Journal,* **4**., Available: *http://www.malariajournal.com/content/4/1/4* [accessed 9 July 2007].

Schmidt, K. A., and Ostfeld, R. S. (2001). Biodiversity and the dilution effect in disease ecology. *Ecology,* **82**, 609-619.

Schneider, P., Takken, W., and McCall, P. J. (2000). Interspecific competition between sibling species larvae of *Anopheles arabiensis* and *An. gambiae. Medical and Veterinary Entomology,* **14**, 165-170.

Service, M. W. (1991). Agricultural development and arthropod borne diseases: a review. *Revista De Saude Publica,* **25**, 165-178.

Service, M. W. (1995). Mosquitoes (Culicidae). In Lane, R. P. and Crosskey, R. W. (Eds.), *Medical Insects and Arachnids.* (2nd edition, pp. 120-240). London, UK: Chapman and Hall.

Service, M. W. (2000). *Medical Entomology for Students.* (2nd edition). Cambridge, UK: Cambridge University Press.

Simberloff, D., and Von Holle, B. (1999). Positive interactions of nonindigenous species: Invasional meltdown? *Biological Invasions,* **1**, 21-32.

Snell, A. E., Derraik, J. G. B., and McIntyre, M. (2005). *Maorigoeldia argyropus* Walker (Diptera: Culicidae): is this another threatened endemic species? *New Zealand Entomologist,* **28**, 95-99.

Solomon, T., and Mallewa, M. (2001). Dengue and other emerging flaviviruses. *Journal of Infection,* **42**, 104-115.

Soper, F. L., and Wilson, D. B. (1943). *Anopheles gambiae* in Brazil 1930-1940.

Sota, T., and Mogi, M. (1989). Effectiveness of zooprophylasxis in malaria control: A throrectical inquiry, with a model for mosquito populations with two bloodmeal hosts. *Medical and Veterinary Entomology,* **3**, 337-345.

Spurr, E., and Sandlant, G. R. (2004). *Risk Assessment for the Establishment of West Nile Virus in New Zealand. Landcare Research Series* No. **25**. Christchurch, NZ, Landcare Research: 27.

Statistics New Zealand. Land use. [online]. (2004). Available from: *http://www.stats.govt.nz/products-and-services/new-zealand-in-profile-2006/environment.htm* [accessed 15 June 2007].

Stenhouse, R. N. (2004). Fragmentation and internal disturbance of native vegetation reserves in the Perth metropolitan area, Western Australia. *Landscape and Urban Planning,* **68**, 389-401.

The World Bank (1997). *World Development Indicators.* Washington DC, USA, The World Bank: 238.

The World Bank (2007). *World Development Indicators.* Washington DC, USA. Available: *http://www.worldbank.org/* [accessed 5 June 2007], The World Bank.

Thompson, R., Begtrup, K., Cuamba, N., Dgedge, M., Mendis, C., Gamagemendis, A., Enosse, S. M., Barreto, J., Sinden, R. E., and Hogh, B. (1997). The Matola Malaria Project: A temporal and spatial study of malaria transmission and disease in a suburban area of Maputo, Mozambique. *American Journal of Tropical Medicine and Hygiene,* **57**, 550-559.

Turell, M. J., Dohm, D. J., Sardelis, M. R., O'Guinn, M. L., Andreadis, T. G., and Blow, J. A. (2005). An update on the potential of North American mosquitoes (Diptera: Culicidae) to transmit West Nile virus. *Journal of Medical Entomology,* **42**, 57-62.

Turell, M. J., and Kay, B. H. (1998). Susceptibility of selected strains of Australian mosquitoes (Diptera: Culicidae) to Rift Valley fever virus. *Journal of Medical Entomology,* **35**, 132-135.

Tyler, K. L. (2001). West Nile Virus Encephalitis in America. *New England Journal of Medicine,* **344**, 1858-1859.

Tyler, K. L. (2004). West Nile virus infection in the United States. *Archives of Neurology,* **61**, 1190-1195.

UNEP (1999). *Global Environmental Outlook* **2000**. Geneva, Switzerland, United Nations Environment Programme.

United Nations Population Division (2006). *World Urbanization Prospects: The 2005 Revision*. New York, NY, United Nations.

USEPA (2000). *Manual for Constructed Wetlands Treatment of Municipal Wastewaters*. Cincinnati, OH, United States Environment Protection Agency.

van den Hurk, A. F., Nisbet, D. J., Hall, R. A., Kay, B. H., MacKenzie, J. S., and Ritchie, S. A. (2003). Vector competence of Australian mosquitoes (Diptera: Culicidae) for Japanese encephalitis virus. *Journal of Medical Entomology,* **40**, 82-90.

VanLeeuwen, J. A., Waltner-Toews, D., Abernathy, T., and Smit, B. (1999). Evolving models of human health toward an ecosystem context. *Ecosystem Health,* **5**, 204-219.

Vezzani, D. (2007). Artificial container-breeding mosquitoes and cemeteries: A perfect match. *Tropical Medicine and International Health,* **12**, 299-313.

Vittor, A. Y., Gilman, R. H., Tielsch, J., Glass, G., Shields, T., Lozano, W. S., Pinedo-Cancino, V., and Patz, J. A. (2006). The effect of deforestation on the human-biting rate of *Anopheles darlingi*, the primary vector of falciparum malaria in the Peruvian Amazon. *American Journal of Tropical Medicine and Hygiene,* **74**, 3-11.

Wackernagel, M., and Rees, W. (1996). *Our Ecological Footprint: Reducing Human Impact on the Earth*. Philadephia, PA: New Society Publishers.

Watson, T. M. (1998). *Ecology and behaviour of Aedes notoscriptus (Skuse): Implications for arbovirus transmission in southeast Queensland*. Brisbane, Australia, University of Queensland. PhD thesis.

Watson, T. M., and Kay, B. H. (1999). Vector competence of *Aedes notoscriptus* (Diptera: Culicidae) for Barmah Forest virus and of this species and *Aedes aegypti* (Diptera: Culicidae) for dengue 1-4 viruses in Queensland, Australia. *Journal of Medical Entomology,* **36**, 508-514.

Weinstein, P. (1994). *The Real and Potential Risks of Human Arboviral Disease in New Zealand*. Wellington, NZ, Communicable Disease Centre, Institute of Environmental Science and Research Ltd.

Weinstein, P. (1995). Arbovirology for New Zealand practitioners. *New Zealand Public Health Report,* **2**, 33-35.

Weinstein, P. (1997). An ecological approach to public health intervention: Ross River Virus in Australia. *Environmental Health Perspectives,* **105**, 364-366.

Weinstein, P. (2005). Human health is harmed by ecosystem degradation, but does intervention improve it? A research challenge from the Millennium Ecosystem Assessment. *EcoHealth,* **2**, 228-230.

Weinstein, P., Laird, M., and Browne, G. (1997). *Exotic and Endemic Mosquitoes in New Zealand as Potential Arbovirus Vectors*. Occasional paper. Wellington, N.Z. Available: *http://www.moh.govt.nz/moh.nsf/wpg_Index/Publications-Online+Publications+Contents* [accessed 17 June 2008]. Ministry of Health.

Weinstein, P., Laird, M., and Calder, L. (1995). Australian arboviruses: At what risk New Zealand? *Australian Journal of Medicine,* **25**, 666-669.

Weinstein, P., Worswick, D., Macintyre, A., and Cameron, S. (1994). Human Sentinels for Arbovirus Surveillance and regional risk classification in South Australia. *The Medical Journal of Australia,* **160**, 494-499.

Weiss, R. A., and McMichael, A. J. (2004). Social and environmental risk factors in the emergence of infectious diseases. *Nature Medicine,* **10**, S70-76.

Wetzel, R. G. (1995). Death, detritus, and energy flow in aquatic ecosystems. *Freshwater Biology,* **33**, 83-89.

Wetzel, R. G., Hatcher, P. G., and Bianchi, T. S. (1995). Natural photolysis by ultraviolet irradiance of recalcitrant dissolved organic matter to simple substrates for rapid bacterial metabolism. *Limnology and Oceanography,* **40**, 1369-1380.

WHO (1982). *Mosquitoes, Mosquito-borne Disease and Mosquito Control Methods: A Review.* Manual on Environmental Management for Mosquito Control with Special Emphasis on Malaria Vectors. Geneva, Switzerland, World Health Organisation: 11-21.

WHO (1996). *The World Health Report 1996: Fighting Disease, Fostering Development.* Geneva, Switzerland, World Health Organization.

WHO (2002). *Malaria in Africa.* Geneva, Switzerland. Available: *http://www.rbm.who.int/cmc_upload/0/000/015/370/RBMInfosheet_3.htm* [accessed 17 June 2008], World Health Organization.

Wilcox, B. A., and Colwell, R. R. (2005). Emerging and reemerging infectious diseases: Biocomplexity as an interdisciplinary paradigm. *EcoHealth,* **2**, 244-257.

Wilcox, B. A., and Gubler, D. J. (2005). Disease ecology and the global emergence of zoonotic pathogens. *Environmental Health and Preventive Medicine,* **10**, 263-272.

Williamson, M. (1996). *Biological Invasions.* New York, NY: Chapman and Hall.

Wilson, C. L. (2001). Ecology and Infectious Disease. In Aron, J. L. and Patz, J. A. (Eds.), *Ecosystem Change and Public Health.* Baltimore, MA: Johns Hopkins University Press.

Wilson, M. E. (1995a). Infectious diseases: an ecological perspective. *British Medical Journal,* **311**, 1681-1684.

Wilson, M. E. (1995b). Travel and the emergence of infectious diseases. *Emerging Infectious Disease,* **1**, 39-46.

World Bank (1993). *World Development Report 1993: Investing in Health.* Oxford, UK: Oxford University Press.

WRI (1998). *World Resources 1998-1999, Environmental Change and Human Health.* Washington, DC, USA, World Resources Institute.

WRI (2000). *World Resources 2000-2001, People and Ecosystems: The Fraying Web of Life.* Washington DC, USA, World Resource Institute.

WRI (2007). *Earth Trends: Environmental Information.* Washington DC, USA. Available: *http://earthtrends.wri.org* [accessed 7 June 2007], World Resources Institute.

In: Focus on Urbanization Trends
Editor: Laura M. De Smet
ISBN 978-1-60456-109-8

Chapter 2

URBANIZATION, ECONOMIC DEVELOPMENT, AND INCOME INEQUALITY: A DEMOGRAPHIC PERSPECTIVE

Yasuhiro Sato[1]*and *Kazuhiro Yamamoto*[2]†
[1]Graduate School of Environmental Studies, Nagoya University,
Fro-cho, Chikusa-ku, Nagoya 464-8601, Japan
[2]Graduate School of Economics, Osaka University,
1-7, Machikaneyama, Toyonaka, Osaka 560-0043, Japan

Abstract

This paper presents a model of endogenous fertility having two sectors (urban and rural) and analyzes the relationship among urbanization, fertility rate determination, and economic development. In the model, technological progress and human capital accumulation are complementary. The urban sector is assumed to have better opportunities for education and a more human capital-intensive production technology than the rural sector. A poverty trap is shown to exist, in which the economy has no human capital accumulation and no technological progress. Once the economy takes off from the poverty trap, human capital accumulation and technological progress advance, and the economy reaches an endogenous growth path. In order for the economy to take off from the poverty trap and begin economic development, sufficiently high human capital and technology levels are shown to be necessary. In addition, the model can replicate one observed pattern of economic development: with economic development, urbanization progresses, fertility declines, and income inequality first rises and then falls.

1. Introduction

It is well known that demographic characteristics play significant roles in economic development. [1] The most prominent phenomenon is that the fertility rate declines as an economy

*E-mail address: ysato@info.human.nagoya-u.ac.jp
†E-mail address: yamamoto@econ.osaka-u.ac.jp
[1]See Ehrlich and Lui (1997) and Galor (2004) for a survey on this subject.

develops. Wrigley (1969) reported that, over the last 200 years, the fertility rate has decreased in many developed countries. In Galor (2004), it was reported that the fertility rate has decreased in European countries since 1870. In addition, he also showed that the fertility rate has decreased in all regions (Western Offshoots, Europe, Africa, Asia, Latin America, and Oceania) since 1960. These facts imply that the fertility rate has decreased with economic development in many countries.

Meanwhile, it is also well known that many countries have experienced advances in the urbanization in the process of economic development. Indeed, the emergence of large cities is a fact that Kuznets (1966) associated with economic development. In addition, the strong positive correlation between economic development and urbanization has been mentioned by economic historians. For example, Bairoch (1988) and Hohenberg and Lees (1985) provided evidence that urbanization had significantly advanced during the period of economic development in Europe. Existing studies, including those by Fujita and Thisse (2002) and Martin and Ottaviano (2001), gave a theoretical explanation of the role of urbanization in the process of economic development using models of an exogenous population.

The purpose of this paper is to analyze the relationship between the above two well-known facts. Thus far, theoretical analysis of this relationship has not attracted much attention from researchers. To the best of our knowledge, Zhang (2002) and Sato and Yamamoto (2005a) are the only exceptions.[2] Sato and Yamamoto (2005a) analyzed how urbanization affects demographic transition via the merits and demerits of population concentration. However, since their model included no engine of endogenous economic growth, they could not examine the relationship between urbanization and the endogenous growth path. Zhang (2002) constructed a two-sector (urban and rural) overlapping-generations model involving the endogenous determination of fertility and education (human capital accumulation), which is one of the best-known engines of endogenous growth. He showed that, when the urban sector has better opportunities for earnings and education than the rural sector, urbanization proceeds, leading to lower fertility as the economy grows endogenously.

In this paper, the two-sector overlapping-generations model described in Zhang (2002) is extended to include technological progress, which is well known as another engine of endogenous growth. In the model, human capital accumulation and technological progress, which are complementary, enable the economy to develop. Furthermore, fertility is determined endogenously: each parent allocates his/her fixed time to working, parenting, and educating children. The incentive for parents to have and educate children originates in the idea that the children will provide material support to the parents in their old age, which is rather a relevant concept in developing countries. The level of children's material support rises as the quantity (number) and quality (educational level) of his/her children increase. Furthermore, the time available for parenting and educating children is limited. Therefore, there is a traditional trade-off in the number of children born and the level of education that they will attain (the so-called "quantity-quality trade-off").[3] The key assumptions here are

[2]There are two empirical studies on this subject: Galor (2004) presented stylized facts that imply (i) that urbanization and economic development started simultaneously and (ii) that, in the early phases of urbanization and economic development, the population growth rate increased, declining in succeeding years. Moreover, Schultz (1985) showed empirically that an advance in the urban sectorization reduces fertility.

[3]Existing studies have provided a rationale for the fertility decline that occurs after an economy takes off from the viewpoint of the quantity-quality trade-off of children. For example, Becker, Murphy, and Tamura

that the urban sector has a more human capital-intensive production technology and more efficient educational technology than does the rural sector. In this context, we examine the influence of educational technology, human capital endowments, and the technological level of the economy on urbanization and fertility determination in the process of economic development. Furthermore, we also investigate the implications on the transition of income inequality.

The main differences between our model and the model described in Zhang (2002) are as follows. First, the motivation of parents to have and educate children. Whereas parents derive utility from the number and educational level of their children in Zhang (2002), parents have a non-altruistic utility function in our model. Second, though there are no migration costs in our model, there are some positive rural-to-urban migration costs in Zhang (2002). Finally, in our model, there is technological progress that is absent from Zhang (2002). The first two differences make the analysis very simple. The third difference generates new qualitative results, as shown below.

The analysis shows that our model has three types of equilibria: a poverty-trap equilibrium, a partial-urbanization equilibrium, and a full-urbanization equilibrium. This is in sharp contrast to Zhang (2002), in which no poverty-trap equilibrium exists. In the poverty-trap equilibrium, no individual invests in the education of his children, and there is no human capital accumulation and no technological progress; thus, the per capita income is constant at the initial level. The initial endowment of human capital determines the level of income inequality.

In the partial-urbanization equilibrium, some individuals live in the rural sector and do not invest in their children's education, whereas other individuals live in the urban sector and invest in their children's education. In this equilibrium, the fertility rates are higher and income is lower in the rural sector than in the urban sector. This income inequality is enlarged over time because human capital accumulation occurs only in the urban sector. Due to human capital accumulation in the urban sector, the economy grows endogenously.

In the full-urbanization equilibrium, all individuals live in the urban sector and invest in their children's education. Technology and human capital accumulation progress infinitely, which implies that the economy is on an endogenous growth path. The level of income inequality becomes constant at the initial level. This is because all people invest in education and the income inequality between parents is taken over by their children, as in the results regarding cross-country income inequality obtained in the standard endogenous growth models.

If the economy has multiple equilibria, the initial levels of technology and human capi-

(1990) constructed a model in which fertility determination is closely related to human capital accumulation. In their model, human capital accumulation raises the returns from working, suggesting a rise in the opportunity costs of having children and also increasing the benefits of educating children. In their model, there are multiple equilibria: a high fertility rate and a low human capital equilibrium and a low fertility rate and high human capital equilibrium. This implies that economic development due to human capital accumulation causes the quantity-quality trade-off of children. Many papers, including those by Kalemli-Ozcan (2002) and Dahan and Tsiddon (1998), have developed models with this type of quantity-quality trade-off.

Technological progress also brings about the quantity-quality trade-off. Technological progress causes a rise in the returns from educating children since education and technological progress are complementary. Then, technological progress affects the relative returns between quantity and quality of children, ultimately causing a demographic transition. Galor and Weil (2000) and Tamura (2002) have presented models of this type.

tal determine the equilibrium that the economy reaches. With these three types of equilibria, we can discuss how to take off from the poverty trap and how the economy develops after the take-off. Moreover, our model can replicate the following features of economic development: urbanization, fertility declines, and rises in income inequality followed by declines in it. Especially, such dynamics in income inequality are not evident in the Zhang (2002) model.

Galor (2004) reported that, in many countries, the economy was once stuck in a poverty trap, in which per capita income is constant and low and the fertility and mortality rates are constant and high.[4] In succeeding years, the economy in many countries started to develop. In this process, urbanization, fertility declines, and high growth of the per capita income progressed. In our model, not only the initial level of technology but also the initial level of human capital must be sufficiently high for the economy to take off from the poverty trap. In addition, if the economy is in a pure-agricultural poverty trap, educational technology in the rural sector plays a significant role: making it sufficiently efficient enables the economy to begin to take off.

Thus far, the relationship between the poverty trap and urbanization has been discussed by Bertinelli and Black (2004). In their two-sector model, human capital investment is possible only in the urban sector, in which individuals always invest in human capital, each individual invests in his/her own human capital, and the population is exogenous. In addition, the level of human capital is determined by the level of technology. The poverty trap was shown to exist, and it was also shown that urbanization can foster economic growth. In contrast, in our model, human capital investment is possible in both sectors, each individual invests in his/her children's human capital, and the population is endogenous. The level of human capital is determined not only by the level of technology but also by the parental level of human capital. These differences generate the following differences in results. In our model, there exists a pure-agricultural poverty trap, in which all the individuals stay in the rural sector, with many children and without human capital accumulation, in order to obtain high current income even though they invest in their children's human capital and economic growth begins if they migrate to the urban sector. Stated differently, individuals obtain current income at the price of future economic growth. This type of poverty trap cannot be observed in the model described in Bertinelli and Black (2004), since human capital accumulation is always impossible without technological progress. Furthermore, while, in our model, economic growth is possible from the pure-agricultural economy, it is impossible in their model. Hence, this study and that by Bertinelli and Black (2004) are complementary.

Our model has various patterns of economic development process. In one pattern, income inequality first rises and then declines, which is confirmed by Barro (2000).[5] Assume that there are two groups of individuals: each individual in one group has high initial endowments of human capital, and each one in the other group has low initial endowments of human capital. Let us consider an economy in the poverty-trap equilibrium in which

[4]In this paper, we do not study the effects of changes in the mortality rate. See Kalemli-Ozcan (2002) and Sato and Yamamoto (2005a) for this point.

[5]However, Barro (2000) admits that this relationship explains only small variations in inequality across countries over time. See Lindeart and Williamson (1985) for the variation of inequality during the period of the Industrial Revolution.

all the individuals stay in the rural sector without human capital accumulation even though they invest in their children's education if they migrate to the urban sector. In this stage, income distribution is constant and is solely determined by the initial endowments of human capital. Now, let us suppose that the economy experiences technological changes, such as increases in complementarity between human capital and technology or improvements in the urban sector production technology. Such changes cause people with high human capital endowments to migrate from the rural to the urban sector, where those people educate their children. Since rural people do not educate their children, the level of income inequality between rural people and urban people increases. As human capital accumulates, it becomes more profitable for individuals to educate their children. This makes migration to the urban sector attractive for them, since education technology is more efficient in the urban sector than in the rural sector. If this force is not very strong, people with low human capital endowments stay in the rural sector, and the economy reaches a partial-urbanization equilibrium, in which human capital is accumulated only by urban people. In this case, increases in the level of income inequality are permanent. If this force is sufficiently strong, people with low human capital endowments also go to the urban sector, and the economy reaches a full-urbanization equilibrium. After this, human capital is accumulated by individuals of both groups. When the economy goes into a full-urbanization equilibrium, income inequality declines because individuals who once did not invest in education begin to educate their children. Thus, our model provides one explanation about the observed pattern of income inequality.

This paper is organized as follows. In Section 2, the basic structure of the model is introduced. Section 3 presents the conditions of the three types of equilibria. Section 4 discusses the implication of the model and presents one possible story of economic development. Section 5 concludes.

2. The Model

Individuals

Consider an economy that consists of two sectors: urban and rural. Relevant variables of the urban carry the subscript u, and those of the rural carry the subscript r. Time is discrete, and each individual lives for three periods: childhood, a young (working) period, and a retirement period. The young period is the sole productive period. At the beginning of the young period, each individual decides (i) where to live, r or u, (ii) the number of children (n_t), (iii) the educational level that each child will attain (e_t), and (iv) the level of consumption they will have (c_t and c_{t+1}). Children live with their parents in the same region, and the elderly live in the region that they chose in their young period. Individual preferences are assumed to take a simple log-linear utility functional form as in Morand (1999), and individuals receive utility from consumption in their young and old periods. [6]

The utility function of time t young individual indexed i is as follows:

[6] In some papers in which the fertility rates are endogenous, such as those by Becker, Murphy, and Tamura (1990), Dahan and Tsiddon (1998), Zhang (2002), and Sato and Yamamoto (2005a), parents have an altruistic preference and derive utility from the number and quality of their children. On the other hand, in Ehrlich and Lui (1991), Morand (1999), and Zhang and Nishimura (1993), the desire for material support during retirement serves as a motive for having children.

$$U_{it} = \ln(c_{it}) + \beta \ln(c_{it+1}). \tag{1}$$

We adopt the utility function used in Ehrlich and Lui (1991) and Morand (1999). In addition, we follow Ehrlich and Lui (1991) and Morand (1999), and assume that each individual has one unit of time in his/her young period and allocates it to rearing and educating children and to working for wages. We assume that childrearing takes b units of time per child and that b is a positive constant. Raising n_{it} identical children and spending e_{it} units of time on education per child, an individual born in period $t-1$ obtains income $w_{it}(1 - bn_{it} - e_{it}n_{it})$, where w_{it} is the wage rate. In this model, we assume that the children are identical. As in Ehrlich and Lui (1991) and Morand (1999), we assume that the capital market is absent and, therefore, children are the parents' only asset available for transferring income from their young to their old period. Our assumption of the nonexistence of a capital market may be extreme. However, as stated in many studies, including those by Goldsmith (1969), Mckinnon (1973), and Levine (1997), capital markets are considered to be highly imperfect in developing countries. Because we focus on developing economies, our assumption would be rational.[7]

In our model, parents bear the (opportunity) costs of parenting and education. In return, they receive a fixed fraction of their children's income in their retirement. It is assumed that agents do not deviate from this behavior. Regarding this point, we follow Ehrlich and Lui (1991), which argued that it can be an equilibrium behavior that children offer their parents a fraction of their income in return for their parents educational investments (implicit contract as they called). Thus, the budget constraint of each individual is as follows:

$$c_{it} + sw_{it}(1 - bn_{it} - e_{it}n_{it}) = w_{it}(1 - bn_{it} - e_{it}n_{it}), \tag{2}$$

$$c_{it+1} = sw_{jt+1}(1 - bn_{jt+1} - e_{jt+1}n_{jt+1})n_{it}, \tag{3}$$

where s is a positive constant satisfying $0 < s < 1$. Each individual spends $100s$ percent of income to support his/her parent. j represents the index of individual i's children. The right-hand side (RHS) of (2) is the young individual's labor income, which is allocated between consumption c_{it} and the support of his/her parent in the form of a fixed fraction s of income, which is shown on the left-hand side (LHS) of (2). (3) indicates that consumption by an old individual is constrained by the amount of support from his/her children. In addition, there is the following time constraint for each young individual:

$$0 \le 1 - bn_{it} - e_{it}n_{it} \le 1. \tag{4}$$

Production and wage structure

Next, we describe the production side of the economy. There are two sectors, urban and rural. We assume that only one kind of consumption goods is produced in the economy. It is treated as the numeraire.

[7]As in the studies by Ehrlich and Lui (1991) and Morand (1999), the results in this paper are qualitatively unaltered even if there are capital markets.

Production technology in the urban sector and that in the rural sector are different. Galor (2004) stated that urban technology is generally more human capital-intensive than rural technology. We follow this argument and set the following wage functions:

$$w_{irt} = G_r A_t h_{it}^{\alpha_r}, \quad (5)$$
$$w_{iut} = G_u A_t h_{it}^{\alpha_u},$$

where A_t represents the economy's level of technology at period t. G_r and G_u are exogenous productivity factors representing, for example, the quality of the rural and urban infrastructure. α_r and α_u represent the human capital intensity in the rural sector production and that in the urban sector production, respectively. α_r and α_u are assumed to be constants satisfying $0 < \alpha_r < \alpha_u < 1$, which implies that the urban technology is more human capital-intensive than rural technology. Equation (5) gives the urban and rural wage rates of an individual i whose human capital level is h_{it}.

Human capital and technological progress

As is common in the literature (e.g., Galor and Tsiddon, 1997), we assume that technological progress is driven by human capital accumulation. Moreover, we assume that human capital accumulation drives technological progress only when it occurs in the urban sector. One reason for this assumption is that innovation activities generally concentrate in the urban sector areas. In fact, Feldman and Florida (1994) observed that innovation activities tend to cluster geographically in areas where research and development (R&D)-oriented firms or universities are established. The economy's level of technology A in period t is initially given at the beginning of period t and A_{t+1} is determined by the following equation of motion:

$$A_{t+1} = \max[1, g(\hat{h}_{ut})]A_t, \quad (6)$$

where $\hat{h}_{ut} = \int_{h_{it} \in H_u} h_{it} dF(h_{it})$ represents the average level of human capital in the urban sector. H_u represents a set of the human capital level of an urban young individual, of which distribution is represented by $F()$. $g(\hat{h}_{ut})$ is a function that governs the features of technological progress. We assume that $g' > 0$, $g'' < 0$, $g(0) < 1$ and $\lim_{\hat{h}_{ut} \to \infty} g(\hat{h}_{ut}) > 1$.

Finally, we formulate human capital accumulation. A young individual i in period t is characterized by a human capital level h_{it}, where h_{it} is distributed according to a cumulative density function $F(h_{it})$ over strictly positive support $[h_t^{min}, h_t^{max}]$. As stated above, each young individual is endowed with one unit of time and allocates it to parenting, educating children, and working. The human capital level of a child h_{jt+1} depends on time of education from his/her parent (e_{it}), on the level of his/her parent's human capital (h_{it}), and on the economy's level of technology at time t (A_t).[8] In addition, we assume that the urban sector has a more efficient educational technology than does the rural sector. This assumption reflects the view that the quality of urban schools is higher than that of rural schools or that urban families can access educational facilities more easily than rural families can.[9]

[8]In Galor and Tsiddon (1997), the human capital level of a child depends on the level of educational investment, on the level of a parent's human capital, and on the technological level. Moreover, the technological level and human capital accumulation are complementary. Our assumption of human capital accumulation follows them.

[9]Zhang (2002) also assumes that an urban sector has a more efficient educational technology than does a rural sector.

More specifically, we suppose that each individual's human capital in each sector is determined as follows:

$$h_{jut+1} = h_{iut}(1+\phi(A_t)e_{iut})^{\gamma}, \tag{7}$$

$$h_{jrt+1} = h_{irt}(1+\theta\phi(A_t)e_{irt})^{\gamma}, \tag{8}$$

where $\phi(A_t) > 0$, $\phi' > 0$, $\phi'' < 0$, $\phi(0) = \underline{\phi} > 0$, $\lim_{A_t \to \infty} \phi(A_t) = \hat{\phi} > 0$. γ is a constant that satisfies $0 < \gamma < 1$. Equations (7) and (8) represent human capital accumulation in the urban sector and that in the rural sector, respectively. Here, $0 < \theta < 1$ represents the level of rural educational technology. This implies that the rural sector has a less efficient educational technology than the urban sector. From (7) and (8), if $e_{it} = 0$, then $h_{jt+1} = h_{it}$; that is, if parents do not invest in the education of their children, the level of their children's human capital remains the same as the level of the parents' human capital.(This holds true both in the urban and rural sectors.)

3. Equilibrium

Fertility rate and educational investment

In this section, we analyze the model and derive an equilibrium. We substitute (2), (3), and (7) into (1) and differentiate it with respect to n_{iut} and e_{iut} to derive first-order conditions. Solving these first-order conditions gives the following fertility rate and level of educational investments in the urban sector:

$$n_{iut} = \frac{\beta(1-\gamma\alpha_u)\phi(A_t)}{(1+\beta)(\phi(A_t)b-1)}, \tag{9}$$

$$e_{iut} = \frac{b\gamma\alpha_u\phi(A_t)-1}{(1-\gamma\alpha_u)\phi(A_t)}. \tag{10}$$

Here, we assume that $\underline{\phi}b > 1$ and $b\gamma\alpha_u\hat{\phi} > 1$. Without the former assumption, the economy collapses because no individual has children when the technological level is low. Without the latter assumption, parents do not invest in the education of their children, even if the technological level reaches its maximum. Since these cases are out of the scope of this paper, we make the above two assumptions.

Substituting (2), (3), and (8) into (1), deriving first-order conditions, and solving them give the fertility rate and the level of educational investments in the rural sector:

$$n_{irt} = \frac{\beta(1-\gamma\alpha_r)\phi(A_t)\theta}{(1+\beta)(\phi(A_t)\theta b-1)}, \tag{11}$$

$$e_{irt} = \frac{b\gamma\alpha_r\phi(A_t)\theta-1}{(1-\gamma\alpha_r)\phi(A_t)\theta}. \tag{12}$$

As seen in (9), (10), (11) and (12), the first-order conditions do not guarantee interior solutions. The reason for this is as follows: from (7) and (8), we can see that the return on educational investments increases with the technological level, implying that in the economy of low technological level, returns on education is low. At the same time, there is a minimum return on quantity of children. If the rate of return on education is below the rate of return on quantity of children regardless of the parents' investment in their children's education, then it is optimal for parents to invest only in quantity. Under such a condition, parents thus choose the corner solution $e_{iut} = 0$ and $e_{irt} = 0$. We can see that e_{iut}, e_{irt}, n_{iut}, and n_{irt} do not depend on the individual's human capital endowment h_{it}. All individuals who live in the same region select the same number of children and their educational level. We then delete the subscripts i from e_{iut}, e_{irt}, n_{iut}, and n_{irt} from here. Note here that the technological level A in period t is determined at the beginning of period t. The level of A_t affects the fertility rate and the educational level as follows. From (9) to (12), and $\phi' > 0$, we can see that

$$\frac{\partial n_{ut}}{\partial A_t} < 0, \frac{\partial n_{rt}}{\partial A_t} < 0, \frac{\partial e_{ut}}{\partial A_t} > 0, \frac{\partial e_{rt}}{\partial A_t} > 0.$$

In both sectors, the fertility rate decreases with technological progress, while educational investments increase with it. In our model, there is a quantity-quality trade-off of children. Since human capital accumulation and technological progress are complementary, the benefits obtained from educating children increase relative to the benefits from having many children as technological progress advances.

From (10) and (12), we can see that $e_{ut} = 0$ and $e_{rt} = 0$ if $b\gamma\alpha_u\phi(A_t) - 1 \leq 0$ and $b\gamma\alpha_r\phi(A_t)\theta - 1 \leq 0$. Because $\phi' > 0$, this implies that parents do not invest in the education of their children if the technological level of the economy is low. Furthermore, in (12), there is a parameter θ that represents the level of rural educational technology. If θ is very low, that is, if education in the rural sector is highly inefficient, parents in the rural sector do not invest in their children's education. Here, we define $\underline{A_u}$ and $\underline{A_r}$ as:

$$\begin{aligned} \underline{A_u} &= \phi^{-1}\left(\frac{1}{b\gamma\alpha_u}\right), \\ \underline{A_r} &= \phi^{-1}\left(\frac{1}{b\gamma\alpha_r\theta}\right). \end{aligned} \quad (13)$$

Lemma 1 *If $A_t \leq \underline{A_u}$, parents do not invest in the education of their children in the urban sector. If $A_t \leq \underline{A_r}$, parents do not invest in the education of their children in the rural sector.*

If $\underline{A_u} < A_t \leq \underline{A_r}$, we have $e_{ut} > 0$ and $e_{rt} = 0$. In this case, educational investments are made in the urban sector but are absent from the rural sector. This is possible because the level of educational technology is lower in the rural sector than in the urban sector ($\theta < 1$) and production technology is less human capital-intensive in the rural sector than in the urban sector ($\alpha_r < \alpha_u$). As we will see later, $e_{ut} > 0$ and $e_{rt} = 0$ are observed in a partial-urbanization equilibrium. More generally, from (10) to (13), we can observe that $n_{ut} \leq n_{rt}$ and $e_{ut} \geq e_{rt}$. The fertility rate is always lower in the urban sector than in the rural sector, while educational investments are larger in the urban sector than in the rural sector.

Lemma 2 *$n_{ut} \leq n_{rt}$ and $e_{ut} \geq e_{rt}$.*

This result is consistent with stylized facts that many researchers have pointed out: the fertility rate is generally lower in urban areas than in rural areas, and educational investments are larger in urban areas than in rural areas.[10]

Locational decision

Next, the locational decision of individuals is studied. As was explained in Section 2, individuals select a sector to live in when they enter the young and working period. They live in a sector in which they can obtain higher utility. Here, it is assumed that, when individuals select a sector to live in, they expect that their children will remain in the same sector when they grow up. This assumption is made for simplicity. Even if parents have perfect foresight, the results are qualitatively unaltered though analysis becomes complicated.[11] Substituting (9) and (10) into (1), and (11) and (12) into (1), we have the following indirect utility functions in the case of $e_{ut} > 0$ and $e_{rt} > 0$, respectively:

$$\begin{aligned} U_{iut} &= \ln\left[\frac{(1-s)G_u A_t h_{it}^{\alpha_u}}{1+\beta}\right] + \beta \ln\left\{ sG_u h_{it}^{\alpha_u}\left[\frac{\alpha_u\gamma(\phi(A_t)b-1)}{1-\alpha_u\gamma}\right]^{\alpha_u\gamma} \right. \\ &\quad \left. \times \frac{\beta(1-\gamma\alpha_u)\phi(A_t)}{(1+\beta)^2(\phi(A_t)b-1)}\right\}, \\ U_{irt} &= \ln\left[\frac{(1-s)G_r A_t h_{it}^{\alpha_r}}{1+\beta}\right] + \beta \ln\left\{ sG_r h_{it}^{\alpha_r}\left[\frac{\alpha_r\gamma(\phi(A_t)b\theta-1)}{1-\alpha_r\gamma\theta}\right]^{\alpha_r\gamma} \right. \\ &\quad \left. \times \frac{\beta(1-\gamma\alpha_r)\phi(A_t)\theta}{(1+\beta)^2(\phi(A_t)b\theta-1)}\right\}. \end{aligned} \tag{14}$$

If $e_{ut} = 0$ and $e_{rt} = 0$, indirect utilities become

$$U_{iut} = \ln\left[\frac{(1-s)G_u A_t h_{it}^{\alpha_u}}{1+\beta}\right] + \beta\ln\left[sG_u h_{it}^{\alpha_u}\frac{\beta}{(1+\beta)^2 b}\right],$$

$$U_{irt} = \ln\left[\frac{(1-s)G_r A_t h_{it}^{\alpha_r}}{1+\beta}\right] + \beta\ln\left[sG_r h_{it}^{\alpha_r}\frac{\beta}{(1+\beta)^2 b}\right].$$

Notice here that U_{iut} and U_{irt} depend on the human capital level h_{it} whereas n_{ut}, n_{rt}, e_{ut}, and e_{rt} don't depend on h_{it}. Therefore, comparing U_{iut} and U_{irt}, we can explore conditions under which an individual having h_{it} live in the urban sector. Because we know that $e_{ut} > e_{rt}$ (from Lemma 2), two possibilities are considered: (i) $e_{ut} > e_{rt} > 0$, and (ii) $e_{ut} > 0$ and $e_{rt} = 0$.

Consider the case of $e_{rt} > 0$ and $e_{ut} > 0$. If $U_{iut} \geq U_{irt}$, an individual having h_{it} lives in the urban sector. From (14), we have

$$U_{iut} \geq U_{irt} \leftrightarrow \left(\frac{G_r}{G_u}\right)^{1-\beta} \leq \Gamma(A_t) h_{it}^{2(\alpha_u-\alpha_r)},$$

where $\Gamma(A_t)$ is defined as

$$\Gamma(A_t) = (\alpha_u\gamma)^{\alpha_u\gamma}\left(\frac{1-\gamma\alpha_u}{\phi(A_t)b-1}\right)^{1-\alpha_u\gamma}\left\{(\alpha_r\gamma)^{\alpha_r\gamma}\left[\frac{(1-\gamma\alpha_r)\theta}{\phi(A_t)b\theta-1}\right]^{1-\alpha_r\gamma}\right\}^{-1}.$$

---

[10] Sharlin(1986) showed that, in the last 100 years, the urban fertility rate has been lower than the rural fertility rate. Sato (2003) showed that an agglomeration economy and a congestion diseconomy in the urban sector lead to low fertility in the urban sector.

[11] A description of the case of parents with perfect foresight is given in Sato and Yamamoto (2005b).

Remind that G_u and G_r are the productivity parameters, and that α_u and α_r represent the human capital intensity in production. From $\phi' > 0$, $0 < \theta < 1$, and $\alpha_u > \alpha_r$, we obtain $\partial\Gamma(A_t)/\partial A_t > 0$. Then, we can understand that technological progress induces urbanization. Since $\Gamma(A_t)h_{it}^{2(\alpha_u-\alpha_r)}$ is an increasing function of h_{it}, it is clear that people who have high human capital live in the urban sector, which implies that human capital accumulation induces urbanization.[12]

In the case of $e_{rt} = 0$ and $e_{ut} > 0$,

$$U_{iut} \geq U_{irt} \leftrightarrow \left(\frac{G_r}{G_u}\right)^{1-\beta} \leq \Lambda(A_t)h_{it}^{2(\alpha_u-\alpha_r)}, \tag{15}$$

where $\Lambda(A_t)$ is defined as

$$\Lambda(A_t) = (\alpha_u\gamma)^{\alpha_u\gamma}\left(\frac{1-\gamma\alpha_u}{\phi(A_t)b-1}\right)^{1-\alpha_u\gamma}\left(\frac{\phi(A_t)}{b}\right).$$

We obtain $\partial\Lambda(A_t)/\partial A_t > 0$.

Lemma 3 *Technological progress and human capital accumulation induce urbanization.*

Three types of equilibrium

With the aid of the three lemmas above, the equilibrium of the model can be examined. Attention is focused on the following three types of equilibria: a poverty-trap equilibrium, a partial-urbanization equilibrium, and a full-urbanization equilibrium. Once the economy goes into one of them, the economy stays in it permanently. If the economy is in an equilibrium other than these three, the economy eventually moves out of it.

In a poverty-trap equilibrium, not all individuals invest in the education of their children. In a partial-urbanization equilibrium, some individuals live in the rural sector and do not invest in their children's education, whereas others live in the urban sector and invest in education. In a full-urbanization equilibrium, all individuals live in the urban sector and invest in their children's education. While human capital accumulation and technological progress do not occur in the first equilibrium, they occur, and the economy grows endogenously, in the latter two equilibria.

Furthermore, if the economy has multiple equilibria, the economy's initial level of technology A_0 and the initial distribution of human capital endowment $F(h_{i0})$ determine the equilibrium that the economy reaches.[13] We assume that $\beta(1-\gamma\alpha_u)\hat{\phi}/[(1+\beta)(\hat{\phi}b-1)] \geq 1$. Without this assumption, population converges to zero in a full-urbanization equilibrium. It is also noteworthy that $n_{rt} \geq n_{ut} \geq 1$ is always satisfied under this assumption.

[12] Funkhouser (1998) showed that migrants from rural to urban areas tend to be more educated than those who do not migrate.

[13] In Zhang (2002), the poverty-trap equilibrium and the partial-urbanization equilibrium do not exist. In Bertinelli and Black (2004), there are poverty-trap, partial-urbanization, and full-urbanization equilibria. However, in their model, people who live in the urban sector areas always invest in human capital, and people who live in the rural sector areas do not always invest in human capital. In addition, in their model, fertility is exogenous. Our model is complementary to models described in Zhang (2002) and Bertinelli and Black (2004).

Poverty-trap equilibrium

In the remainder of this section, the conditions under which the above three types of equilibria are attained are investigated. First, the conditions under which the economy is in a poverty-trap equilibrium are derived. As seen in Section 2, if the following inequality holds, human capital accumulation does not occur in either sector:

$$A_0 \leq \underline{A_u}. \tag{16}$$

If this inequality holds, parents do not invest in their children's education in the urban sector. Moreover, (13) implies that the inequality $A_0 \leq \underline{A_r}$ also holds true when (16) is satisfied. Hence, individuals do not invest in their children's education in the rural sector. From this argument, it is clear that, if the initial technological level of the economy is low, individuals don't invest in education at all.

Here, for the following discussions, h^* and $\bar{h}$, which satisfy the following equations, are defined:

$$\left(\frac{G_r}{G_u}\right)^{1-\beta} = \Lambda(A_0)h^{*2(\alpha_u-\alpha_r)}.$$

$$\left(\frac{G_r}{G_u}\right)^{1-\beta} = \lim_{A_t\to\infty} \Lambda(A_t)\bar{h}^{2(\alpha_u-\alpha_r)}.$$

From (15), individuals who have $h_0 \geq h^*$ live in the urban sector, while individuals who have $h_0 < h^*$ live in the rural sector at the initial time in the case of $\underline{A_u} < A_0 < \underline{A_r}$. Because $\partial\Lambda/\partial A_t > 0$, we have $\bar{h} \leq h^*$. If $h_0^{min} < \bar{h}$ and $b\gamma\alpha_r\hat{\phi}\theta \leq 1$, some individuals live in the rural sector and do not invest in education.

When (16) does not hold, individuals educate their children in the urban sector. However, if all the individuals live in the rural sector and if rural individuals do not invest in education, human capital accumulation does not occur. This situation is described by the following three inequalities:

$$A_0 > \underline{A_u},$$
$$A_0 \leq \underline{A_r},$$
$$h_0^{max} < h^*.$$

$A_0 > \underline{A_u}$ implies that individuals invest in education if they live in the urban sector. However, since $A_0 \leq \underline{A_r}$ and $h_0^{max} < h^*$ are satisfied, all individuals live in the rural sector, and rural individuals do not invest in education. Thus, if these three inequalities are satisfied, human capital accumulation does not occur at the initial time.

We saw that, if $g(\hat{h}_{ut}) \leq 1$, the economy's level of technology is constant. Then, if $g(\hat{h}_{u0}) \leq 1$ (i.e., the initial endowment of human capital in the whole economy is low) and no individual invests in education, technological progress never occurs in this economy.

From the above arguments, the two sets of conditions for a poverty-trap equilibrium can be derived. Both sets include conditions of no human capital accumulation and no technological progress. One consists of the following two conditions: $A_0 \leq \underline{A_u}$ and $g(\hat{h}_{u0}) \leq 1$. In this case, no one invests in education, and technological progress does not occur. The other consists of the following two conditions: $h_0^{max} < h^*$ and $A_0 \leq \underline{A_r}$. In this case, all people live in the rural sector, and no one invests in education.

Proposition 1 *If the economy satisfies one of the following two sets of conditions, the economy is in the poverty-trap equilibrium.*

- $A_0 \leq \underline{A_u}$ *and* $g(\hat{h}_{u0}) \leq 1$.
- $h_0^{max} < h^*$ *and* $A_0 \leq \underline{A_r}$.

Partial-urbanization equilibrium

Next, the conditions under which the economy attains a partial-urbanization equilibrium are derived. In this equilibrium, both sectors have residents. Urban individuals invest in their children's education, while rural individuals do not. The condition under which human capital accumulation occurs in the urban sector at the initial time is that $A_0 > \underline{A_u}$. Because human capital accumulation occurs in the urban sector, $g(\hat{h}_{ut})$ increases to be greater than 1 in time since $\lim_{\hat{h}_{ut} \to \infty} g(\hat{h}_{ut}) > 1$. Therefore, technological progress occurs in the economy and $\lim_{t \to \infty} A_t = \infty$. With this technological progress, $\lim_{t \to \infty} \phi(A_t) = \lim_{A_t \to \infty} \phi(A_t) = \hat{\phi}$ in this equilibrium. Because of human capital accumulation in the urban sector and technological progress, the economy is in an endogenous growth process.

The condition under which both sectors have residents is $h_0^{min} < h^* < h_0^{max}$. In a partial-urbanization equilibrium, rural individuals do not educate their children. In order for the economy to stay in this equilibrium, $h_0^{min} < \bar{h}$ and $b\gamma\alpha_r\hat{\phi}\theta \leq 1$ are necessary. The former ensures that some individuals live in the rural sector permanently, even if technological progress continues in the economy, while the latter ensures that rural individuals do not invest in education permanently due to low rural educational technology.

Proposition 2 *The economy is in a partial-urbanization equilibrium if the following conditions are satisfied:*

- $h_0^{min} < \bar{h} \leq h^* < h_0^{max}$, $b\gamma\alpha_r\hat{\phi}\theta \leq 1$, *and* $A_0 > \underline{A_u}$.

Moreover, even if the economy is not in a partial-urbanization equilibrium, the economy reaches it over time under weaker conditions. If $g(\hat{h}_{u0}) > 1$, some agents live in the urban sector at the initial time, and technological progress occurs. In this case, there must be a time t at which $A_t = \underline{A_u}$, if $A_0 < \underline{A_u}$. Hence, we have the following proposition.

Lemma 4 *The economy reaches a partial-urbanization equilibrium over time if the following conditions are satisfied:*

- $h_0^{min} < \bar{h}$, $b\gamma\alpha_r\hat{\phi}\theta \leq 1$, *and* $g(\hat{h}_{u0}) > 1$.

Full-urbanization equilibrium

Finally, we derive the conditions for a full-urbanization equilibrium. In this equilibrium, both human capital accumulation and technological progress occur, and the economy follows the endogenous growth process.

It is noteworthy that, in order for the economy to be in the full-urbanization equilibrium, human capital accumulation needs to occur in the urban sector.[14] We derived the

[14]In our model, if human capital accumulation does not occur in the urban sector, human capital accumulation never occurs in the rural sector. See (11) and (13) in Section 2.

condition for this in the discussion regarding the partial-urbanization equilibrium: $A_0 > \underline{A_u}$ or $g(\hat{h}_{u0}) > 1$. If $A_0 > \underline{A_u}$ and the inequality $h^* < h_0^{min}$ hold, all individuals live in the urban sector and invest in education.

Proposition 3 *The economy is in the full-urbanization equilibrium if the following conditions hold:*

- $h^* < h_0^{min}$ *and* $A_0 > \underline{A_u}$.

Moreover, if $A_0 > \underline{A_r}$, human capital accumulation progresses in both regions throughout time. Then, the economy reaches a full-urbanization equilibrium over time.

If $A_0 > \underline{A_u}$ and the inequality $h^* < h_0^{max}$ hold, individuals who live in the urban sector invest in education throughout time, and techonological progress occurs. In this case, if $\overline{h} < h_0^{min}$, the resident whose human capital endowments are h_0^{min} lives in the urban sector at the steady state. Then, in the case of $A_0 > \underline{A_u}$, $h^* < h_0^{max}$, and $h < h_0^{min}$, the economy reaches a full-urbanization equilibrium.

In addition, if $A_0 > \underline{A_u}$ and $h^* < h_0^{max}$ are satisfied, human capital accumulation occurs in the rural sector in the case of $b\gamma\alpha_r\hat{\phi}\theta > 1$. Then, residents can accumulate human capital in the rural sector. In this case, all residents live in the urban sector in the steady state, and human capital accumulation and technological progress occur. Therefore, if $A_0 > \underline{A_u}$, $h^* < h_0^{max}$, and $b\gamma\alpha_r\hat{\phi}\theta > 1$ are satisfied, the economy reaches a full-urbanization equilibrium.

If $g(\hat{h}_{u0}) > 1$, some agents live in the urban sector at the initial period, and technological progress occurs. This technological progress derives the human capital accumulation in the urban sector, even if $A_0 < \underline{A_u}$. Then, with similar arguments of the case of $A_0 > \underline{A_u}$, we can derive the conditions of the full-urbanization equilibrium: $g(\hat{h}_{u0}) > 1$ and $\overline{h} < h_0^{min}$ or $g(\hat{h}_{u0}) > 1$ and $b\gamma\alpha_r\hat{\phi}\theta > 1$.

Lemma 5 *The economy reaches a full-urbanization equilibrium over time if one of following five sets of conditions holds:*

- $A_0 > \underline{A_r}$.
- $A_0 > \underline{A_u}$, $\overline{h} < h_0^{min}$, *and* $h^* < h_0^{max}$.
- $A_0 > \underline{A_u}$, $h^* < h_0^{max}$, *and* $b\gamma\alpha_r\hat{\phi}\theta > 1$.
- $\overline{h} < h_0^{min}$ *and* $g(\hat{h}_{u0}) > 1$.
- $b\gamma\alpha_r\hat{\phi}\theta > 1$ *and* $g(\hat{h}_{u0}) > 1$.

From Propositions 1 to 3, we can understand that, if A_0 and $\hat{h}_{u0}$ are small, the economy is in the poverty-trap equilibrium. If A_0 and $\hat{h}_{u0}$ are high enough, the economy eventually reaches a partial-urbanization equilibrium or a full-urbanization equilibrium.

Thus, we can see that our model has two sets of multiple equilibria: one consists of a poverty-trap equilibrium and a partial-urbanization equilibrium, and the other consists of a poverty-trap equilibrium and a full-urbanization equilibrium.

4. Urbanization, Income Inequality, and Economic Development

In this section, the characteristics of three equilibria, namely, poverty-trap, partial-urbanization, and full-urbanization equilibria, are summarized. One story of economic development that captures some features of economic development is then presented.

In a poverty-trap equilibrium, human capital accumulation and technological progress do not occur, and the fertility rate is constant at $\beta/[(1+\beta)b]$. Individuals who have human capital endowment $h_0 > h^*$ live in the urban sector, whereas individuals with $h_0 < h^*$ live in the rural sector. The level of income inequality is constant at the initial level.

In a partial-urbanization equilibrium, individuals with $h_0 > \bar{h}$ live in the urban sector, whereas individuals with $h_0 < \bar{h}$ live in the rural sector. Although the urban fertility rate decreases over time, the rural fertility rate is constant at $\beta/[(1+\beta)b]$, which is always higher than the urban fertility rate. Human capital accumulation progresses in the urban sector and technological progress occurs, which makes the economy grow endogenously. However, the level of human capital is constant in the rural sector because rural parents do not invest in their children's education. Therefore, income inequality between urban people and rural people increases over time.

In the full-urbanization equilibrium, all people live in the urban sector. The fertility rate decreases, and educational investments increase. Human capital accumulation and technological progress continue, and the economy is on the endogenous growth path. The income inequality is constant at the initial level. This is because all people invest in education and the income inequality between parents is taken over by their children, as in the results regarding cross-country income inequality obtained in the standard endogenous growth models.

With these three equilibria, we can show one story that captures some features of economic development. Consider an economy with $h_0^{max} < h^*$, $A_0 \leq \underline{A_r}$, and $A_0 < \underline{A_u}$. This economy is in a poverty-trap equilibrium (from Proposition 1), and all people live in the rural sector. Furthermore, they do not educate their children. Hence, the fertility rate is constant at its maximum ($n_t = \beta/[(1+\beta)b]$), and there are no human capital accumulation or technological progress. The level of income of each individual depends entirely on the initial endowment of human capital, which is constant over time. Then, the level of income inequality is constant and relatively low. This state may be referred to as the Malthusian state a la Galor (2004).[15][16]

Let us suppose that the economy experiences technological changes and h^* decreases to less than h_0^{max}. This can be caused by rises in intial technology level A_0. When h^* decreases to less than h_0^{max}, individuals with $[h^*, h_0^{max}]$ migrate from the rural to the urban sector, where they begin to invest in education. At this time, human capital accumulation and technolog-

[15]Galor (2004) pointed out that, in the Malthusian epoch, the fertility rate and the infant mortality rate are high and the income level is mired at a low level because there are no human capital accumulation or technological progress.

[16]In this model, we do not study the aspect of the positive infant mortality rate for technical simplicity. The focus is on urbanization, endogenous fertility, and economic development. Kalemi-Ozcan (2002, 2003) and Sato and Yamamoto (2005a) reported the influence of declines in infant mortality rate on economic development with fertility declines.

ical progress occur in the urban sector. As A_t rises, h^* further decreases, and more people move from the rural sector to the urban sector. Migration ends when the economy reaches either a partial-urbanization equilibrium or a full-urbanization equilibrium (Propositions 2 and 3). The economy reaches the former if $h_0^{min} < \bar{h}$ and $b\gamma\alpha_r\hat{\phi}\theta \leq 1$. It reaches the latter otherwise. During this epoch, urbanization progresses and urban individuals educate their children, while rural individuals do not. Therefore, income inequality between urban and rural people increases over time. In this process, the national fertility rate decreases with technological progress and urbanization. From (9) to (12), we can see that technological progress raises educational investments per child and decreases the fertility rate. In addition, urbanization decreases the national fertility rate because the fertility rate is lower in the urban sector than in the rural sector (Lemma 3). This story of economic development is consistent with the economic development in the western Europe described in Galor (2004). In fact, he showed that in the mids of 18th century, England experienced the industrial revolution, which is followed by France and other western European countries during the 19th century. We can interpret this phenomenon as follows: industrial revolution in England raises the technological level A_0 in England, and after this, other western European countries absorbed technology from England and experienced rises in A_0.[17] This enabled each country to be on the growth path, which was accompanied by urbanization.

Galor (2004) also pointed out the important role of public education, which improves the education technology both in rural and urban areas, leading to a significant increase in supply of educated workers in the second half of the 19th century. For example, in England, the proportion of children aged 5 to 14 in primary schools increased from 11 percent in 1855 to 74 percent in 1900. As shown in Flora et. al. (1983), a similar pattern is observed in other European countries. In paticular, in France the perventage increased from 30 percent in 1832 to 86 percent in 1901. If the education technology is developed enough to make $\bar{h}$ lower than h_0^{min}, the economy can reach a full-urbanization equilibrium. This is because advances in urbanization enhance technological progress and raise A_t. Hence, h^* decreases and eventually reaches very close to $\bar{h}$, which implies that $h^* < h_0^{min}$ and that the economy is in a full-urbanization equilibrium. When there becomes no migration, income inequality decreases because individuals who once did not educate children in the rural sector begin to educate children in the urban sector. However, this decrease does not continue, and the income inequality becomes constant at the level of initial time in a full-urbanization equilibrium.

These phenomena (urbanization, fertility declines, and increases and decreases in income inequality) are consistent with observations reported in the Introduction. Of course, our model has another pattern of economic development. When an economy in the poverty-trap equilibrium is considered as reported above, and supposing that improvements are made in the educational technology of the rural sector (increases in θ), then, rural individuals come to invest in education, and economic development begins.

[17]Gerschenkron (1962) pointed out the importance of absorbing technologies from developed countries for developing countries.

5. Concluding Remarks

This paper presented the mechanism of urbanization, fertility decline, and transition of income inequality with economic development. If the initial human capital endowment and the technological level of the economy are low, the economy reaches a poverty-trap equilibrium. In this equilibrium, human capital accumulation and technological progress do not occur, and the fertility rate is constant at a high level. The level of income inequality is constant at the initial level. If the initial human capital endowment and the technological level of the economy are high, the economy reaches either a partial-urbanization equilibrium or a full-urbanization equilibrium. If the level of rural educational technology is low, the economy reaches the former. In this equilibrium, individuals with low human capital endowments live in the rural sector and do not invest in education, while those with high human capital endowments live in the urban sector and educate their children. If the level of rural educational technology is high, the economy reaches a full-urbanization equilibrium. In this equilibrium, all individuals live in the urban sector and invest in education. Fertility decreases over time. Both human capital accumulation and technological progress occur, and the average per capita income grows at its maximum value. With these three types of equilibria, our model showed the following story of economic development: urbanization progresses, fertility decreases, and income inequality first rises and then falls.

Understanding the relationship among urbanization, fertility determination, and economic development is crucial for the interpretation of past and current trends and events. However, only a few studies have investigated this topic thus far. This paper presented one possible mechanism that would help understand the relationship among these factors.

Acknowledgements

We are grateful to Masahisa Fujita, Takatoshi Tabuchi, Tomoya Mori, Zeng Dao-Zhi, Hajime Takatsuka, Tatsuaki Kuroda, Yasusada Murata, Hikaru Ogawa, Makoto Tawada, Ryuhei Okumura, and Yoshitaka Ishikawa for useful comments and suggestions. Any remaining errors are ours. The first author acknowledges the financial support from the Japan Society for the Promotion of Science through Grants-in-Aid for Scientific Research ((B)14330004 and (S)18103002), and a Grant-in-Aid for Young Scientists ((B)18730162).

Appendix
Perfect-Foresight Parents

In this appendix, we consider perfect-foresight parents. We must distinguish two cases: $b\gamma\alpha_r\phi(A_t)\theta > 1$ and $b\gamma\alpha_r\phi(A_t)\theta \leq 1$. In the case of $b\gamma\alpha_r\phi(A_t)\theta > 1$, rural parents invest in their children's education.

Here, we assume that rural parents expect that their children will migrate to the urban sector in the next period. The first-order conditions of these parents are as follows:

$$n_{rut} = \frac{\beta(1-\gamma\alpha_u)\phi(A_t)\theta}{(1+\beta)(\phi(A_t)\theta b - 1)},$$

$$e_{rut} = \frac{b\gamma\alpha_u\phi(A_t)\theta - 1}{(1-\gamma\alpha_u)\phi(A_t)\theta}.$$

From these conditions, we can derive the indirect utilities of these parents,

$$\begin{aligned} U_{irut} &= \ln\left[\frac{(1-s)G_rA_th_{it}^{\alpha_r}}{1+\beta}\right] + \beta\ln\left\{sG_uh_{it}^{\alpha_u}\left[\frac{\alpha_u\gamma(\phi(A_t)b\theta-1)}{1-\alpha_u\gamma\theta}\right]^{\alpha_u\gamma}\right. \\ &\quad \times \left.\frac{\beta(1-\gamma\alpha_u)\phi(A_t)\theta}{(1+\beta)^2(\phi(A_t)b\theta-1)}\right]. \end{aligned}$$

If $U_{rut} \leq U_{ut}$, all parents expect their children to migrate to the urban sector, and the children do so. We define h^m, which satisfies $U_{rut} = U_{ut}$. Some calculations show that

$$h^m = \left(\frac{G_r}{G_u}\right)^{1/(\alpha_u-\alpha_r)} Z(A_t),$$

where $Z(A_t)$ is defined as

$$Z(A_t) = \left[\frac{(1-\alpha_u\gamma)(\phi(A_t)b\theta-1)}{(1-\alpha_u\gamma\theta)(\phi(A_t)b-1)}\right]^{\beta\alpha_u\gamma/(\alpha_u-\alpha_r)} \left[\frac{\theta(\phi(A_t)b-1)}{\phi(A_t)b\theta-1}\right]^{\beta/(\alpha_u-\alpha_r)}.$$

If $h_{it} \geq h^m$, $U_{iut} \geq U_{irut}$. If parents have human capital $h^m \leq h_{it} \leq h^c$ and get the wage $w_{iut} < w_{irt}$, their children must get human capital $h_{jt+1} > h^c$ and the wage $w_{jut+1} > w_{jrt+1}$, since $U_{iut} \geq U_{irut}$ and $n_{rt} > n_{ut}$.

Here, we define h^b as $(G_r/G_u)^{1-\beta} = \Gamma(A_t)h^{b2(\alpha_u-\alpha_r)}$. Then, if $h_{it} > (<) h^b$, $U_{iut} > (<) U_{irt}$ when $e_{rt} > 0$ and $e_{ut} > 0$.

Here, we can identify the following six cases:

1. $U_{irt} < U_{iut} < U_{irut}$
2. $U_{irt} < U_{irut} < U_{iut}$
3. $U_{iut} < U_{irt} < U_{irut}$
4. $U_{iut} < U_{irut} < U_{irt}$
5. $U_{irut} < U_{iut} < U_{irt}$
6. $U_{irut} < U_{irt} < U_{iut}$

It is clear that, in cases 3 and 4, agents live in the rural sector and, in cases 2 and 6, they live in the urban sector. Here, we derive the value of h^c, which satisfies $w_{ut} = w_{rt}$. From (5),

$$h^c = \left(\frac{G_r}{G_u}\right)^{1/(\alpha_u-\alpha_r)}.$$

Moreover, $\partial U_{irut}/\partial\theta > 0$ implies that

$$\begin{aligned} & ln\left\{sG_uh_{it}^{\alpha_u}\left[\frac{\alpha_u\gamma(\phi(A_t)b-1)}{1-\alpha_u\gamma}\right]^{\alpha_u\gamma}\frac{\beta(1-\gamma\alpha_u)\phi(A_t)}{(1+\beta)^2(\phi(A_t)b-1)}\right\} \\ > \; & ln\left\{sG_uh_{it}^{\alpha_u}\left[\frac{\alpha_u\gamma(\phi(A_t)b\theta-1)}{1-\alpha_u\gamma\theta}\right]^{\alpha_u\gamma}\frac{\beta(1-\gamma\alpha_u)\phi(A_t)\theta}{(1+\beta)^2(\phi(A_t)b\theta-1)}\right\}. \end{aligned}$$

Thus, if $h_{it} \geq h^c$ and $U_{irt} < U_{iut}$, $U_{irt} < U_{irut} < U_{iut}$. Thus, agents who have human capital endowments $h_{it} \geq h^c$ live in the urban sector when $U_{irt} < U_{iut}$. In addition, their children also live in the urban sector because they have human capital endowments $h_{jt+1} \geq h_{it} \geq h^c$.

In case 1, if their children live in the urban (rural) sector, the agents live in the rural (urban) sector. Since $U_{irt} < U_{iut}$, their children's human capital endowments h_{jt+1} overcome h^c. Then, from the above discussion, the agents' children live in the urban sector. Therefore, in case 1, agents live in the rural sector. In case 5, if their children live in the rural (urban) sector, the agents live in the rural (urban) sector. In this case, $h^m < h_{it} < h^b$. Here, because $b\gamma\alpha_r\phi(A_t)\theta > 1$, there must be a time $t+x$ when $h_{jt+x} > h^b$. Agents who have $h_{jt+x} > h^b (> h^m)$ live in the urban sector. Thus, their parents, who live in time $t+x-1$, live in the urban sector, and the agents of time $t+x-2$ live in the urban sector. Thus, the agents of $h^m < h_{it} < h^b$ live in the urban sector because their children live in the urban sector. Therefore, in case 5, agents live in the urban sector.

In case 1, $h^b < h^m$, and, in case 5, $h^m < h^b$. If $h^b < h^m$, the agents of $h_{it} \leq h^b$ live in the rural sector, since they are in the case of 3 or 4. On the other hand, if $h^b < h^m$, agents who have a $h_{it} \geq h^m$ live in the urban sector, since they are in the case of 2 or 6. If $h^m < h^b$, the agents of $h_{it} \leq h^m$ live in the rural sector, since they are in the case of 3 or 4. On the other hand, if $h^m < h^b$, agents who have a $h_{it} \geq h^b$ live in the urban sector, since they are in the case of 2 or 6. Then, it is clear that, if $b\gamma\alpha_r\phi(A_t)\theta > 1$, agents with $h_{it} > (<) h^m$ live in the urban (rural) sector.

Next, let us focus on the case of $b\gamma\alpha_u\phi(A_t)\theta > 1$ and $b\gamma\alpha_r\phi(A_t)\theta \leq 1$. In this case, rural parents whose children will live in the urban sector in the next period invest into the education of their children. On the other hand, rural parents whose children will live in the rural sector at the next time do not invest into the education of their children. When $b\gamma\alpha_u\phi(A_t)\theta > 1$ and $b\gamma\alpha_r\phi(A_t)\theta \leq 1$, agents whose $U_{iut} > \max(U_{irut}, U_{irt})$ live in the urban sector, and agents whose $U_{iut} \leq \max(U_{irut}, U_{irt})$ live in the rural sector.

$$h^0 = \left[\left(\frac{G_r}{G_u} \right)^{(1-\beta)} (\alpha_u\gamma)^{\alpha_u\gamma} \left(\frac{1-\gamma\alpha_u}{\phi(A_t)b-1} \right)^{(1-\alpha_u\gamma)} \frac{\phi(A_t)}{b} \right]^{1/[2(\alpha_u-\alpha_r)]}.$$

When $b\gamma\alpha_u\phi(A_t)\theta > 1$ and $b\gamma\alpha_r\phi(A_t)\theta \leq 1$, agents with $h_{it} > \max(h^m, h^0)$ live in the urban sector, and agents with $h_{it} \leq \max(h^m, h^0)$ live in the rural sector.

In the case of $b\gamma\alpha_u\phi(A_t)\theta \leq 1$, children whose parents live in the rural sector also continue to live in the rural sector because, in the rural sector, $h_{jt+1} = h_{it}$. When $b\gamma\alpha_u\phi(A_t)\theta \leq 1$ and $b\gamma\alpha_u\phi(A_t) > 1$, agents with $h_{it} > (\leq) h^0$ live in the urban (rural) sector.

In the case of $b\gamma\alpha_u\phi(A_t) \leq 1$, agents live in the urban (rural) sector if $U_{iut} > (<) U_{irt}$ with $e_{rt} = e_{ut} = 0$. This is the same as in the Discussion in the text.

As we saw in this appendix, agents with high enough human capital endowments live in the urban sector, and agents with low human capital endowments live in the rural sector. These results are qualitatively the same as those in our model, described in the text. We can derive three types of equilibria: a poverty-trap equilibrium, a partial-urbanization equilibrium, and a full-urbanization equilibrium with the assumption of perfect-foresight parents. However, as we saw in this appendix, the migration behavior of perfect-foresight parents becomes more complex than that in our model, although the main point does not change: agents with high human capital live in the urban sector, and those with low human capital live in the rural sector. Then, our assumption is only for the sake of simplification of the analysis.

References

[1] Bairoch, P., (1988) *Cities and Economic Development*, (C. Braider, trans.), Chicago: University of Chicago Press.

[2] Barro, R., (2000) "Inequality and Growth in a Panel of Countries." *Journal of Economic Growth* **5**, 5-32.

[3] Becker, G. S., K. M. Murphy, and T. Tamura, (1990) "Human Capital, Fertility, and Economic Growth." *Journal of Political Economy* **98**, 12-37.

[4] Bertinelli, L., and D. Black, (2004) "Urbanization and Growth." *Journal of Urban Economics* **56**, 80-96. 33, 335-360.

[5] Dahan, M., and D. Tsiddon, (1998) "Demographic Transition, Income Distribution, and Economic Growth." *Journal of Economic Growth* **3**, 29-52.

[6] de la Croix D., and Doepke, M. (2003) Inequality and Growth: Why Differential Fertility Matters? *American Economic Review* **93**: 1091-1113.

[7] de la Croix, D., and Doepke, M. (2004) Public Versus Private Education When Differential Fertility Matters. *Journal of Development Economics* **73**: 607-629.

[8] Ehrlich, I., and F. T. Lui, (1991) "Intergenerational Trade, Longevity, and Economic Growth." *Journal of Political Economy* **99**, 1029-1059.

[9] Ehrlich, I., and F. T. Lui, (1997) "The Problem of Population and Growth: a Review of Literature from Malthus to Contemporary Models of Endogenous Population and Endogenous Growth." *Journal of Economic Dynamics and Control* **21**, 205-242.

[10] Feldman, M. P., and R. Florida, (1994) "The Geographic Sources of Innovation: Technological Infrastructure and Product Innovation in the United States." *Annals of the Association of American Geographers* **84**, 210-229.

[11] Flora P., Kraus F., and Pfenning, F. (1983) *State Economy and Society in Western Europe 1815*∼1675. vol. 1. St James Press.

[12] Fujita, M., and J. F. Thisse, (2002) *Economics of Agglomeration*, Cambridge University Press.

[13] Funkhouser, E., (1998) "Changes and the Returns to Education in Costa Rica." *Journal of Development Economics* **57**, 289-317.

[14] Galor, O., and D. Tsiddon, (1997) "The Distribution of Human Capital and Economic Growth." *Journal of Economic Growth* **2**, 93-124.

[15] Galor, O., and D. N. Weil, (1999) "Population, Technology, and Growth: From Malthusian Stagnation to the Demographic Transition and Beyond." *American Economic Review* **90**, 806-828.

[16] Galor, O., (2004) "From Stagnation to Growth: Unified Growth Theory." in *the Handbook of Economic Growth*, (P. Aghion and S. Durlauf eds.), *forthcoming*.

[17] Gerschenkron, A., (1962) *Economic Backwardness in Historical Perspective.* Cambridge, Mass.: Harvard University Press.

[18] Goldsmith, Raymond, W. (1969). *Financial Structure and Development.* New Haven, Conn.: Yale University Press.

[19] Hohenberg, P., and L. H. Lees, (1985) *The Making of Urban Europe (1000-1950).* Oxford: Oxford University Press.

[20] Iygun, M. F., (2004) "Geography, Demography, and Early Development." *Journal of Population Economics, forthcoming.*

[21] Jacobs, J., (1969) *The Economy of Cities*. New York: Random House.

[22] Kalemli-Ozcan, S., (2002) "Does the Mortality Decline Promote Economic Growth?" *Journal of Economic Growth* **7**, 411-439.

[23] Kremer, M., and D. L. Chen, (2002) "Income Distribution Dynamics with Endogenous Fertility." *Journal of Economic Growth* 7, 227-258.

[24] Kuznets, S., (1966) *Modern Economic Growth, Rate Structure, and Spread.* New Haven: Yale University Press.

[25] Levine, R., (1997) "Financial Development and Economic Growth: Views and Agenda." *Journal of Economic Literature* **35**, 688-726.

[26] Lindert, P. H., and J. G. Williamson, (1985) "Growth, Equality, and History." *Explorations in Economic History* **22**, 341-377.

[27] McKinnon, R., (1973) *Money and Capital in Economic Development.* Washington, D.C.: Brooking Institute.

[28] Martin, P., and G. I. P. Ottaviano, (2001) "Growth and Agglomeration." *International Economic Review* **42**, 947-968.

[29] Morand, O., (1999) "Endogenous Fertility, Income Distribution, and Economic Growth." *Journal of Economic Growth* **4**, 221-230.

[30] Sato, Y., (2007) "Fertility and Economic Geography." *Journal of Urban Economics* **61**, 372-387.

[31] Sato, Y., and K. Yamamoto, (2005a) "Population Concentration, Urbanization, and Demographic Transition," *Journal of Urban Economics* **58**, 45-61.

[32] Sato, Y., and K. Yamamoto (2005b) "Urbanization and Poverty Trap: a Demographic Perspective," *DEE Discussion paper No.05-2, Department of Economic Environment, Graduate School of Environmental Studies, Nagoya University* .

[33] Sharlin, A., (1986) "Urban-Rural Differences in Fertility in Europe during the Demographic Transition." in *The Decline of Fertility in Europe*, (A. J. Coale and S. C. Watkins, eds.), Princeton, NJ: Princeton University Press.

[34] Schultz, T. P., (1985) "Changing World Prices, Women's Wages, and the Fertility Transition: Sweden, 1860-1910." *Journal of Political Economy* **93**, 1126-1154.

[35] Tamura, T., (2002) "Human Capital and the Switch from Agriculture to Industry." *Journal of Economic Dynamics and Control* **27**, 207-242.

[36] Wrigley, E. A., (1969) *Population and History.* New York, NY: McGraw-Hill.

[37] Zhang, J., and K. Nishimura, (1993) "The Old-Age Security Hypothesis Revisited," *Jornal of Development Economics* **41**, 191-202.

[38] Zhang, J., (2002) "Urbanization, Population Transition, and Growth." *Oxford Economic Papers* **54**, 91-117.

In: Focus on Urbanization Trends
Editor: Laura M. De Smet
ISBN: 978-1-60456-109-8

Chapter 3

AMYOTROPHIC LATERAL SCLEROSIS AND URBANIZATION IN THE HEALTH DISTRICT OF FERRARA, ITALY, IN 1964-1998

Vittorio Govoni, Enrico Granieri, Ilaria Casetta and Elisa Fallica*
Neurologic Clinic of the University of Ferrara
*Neurology Unit of the University Hospital of Ferrara, Italy

Abstract

The present chapter develops the issue of the supposed relationship between the incidence of ALS and environmental factors, and in particular the supposed relationship with those environmental factors that are in relation with the level of urbanization or in relation with the rural environment.The hypothesis that changing in the frequency of the disease could be the consequence of environmental changes is supported. This chapter has first considered the available data from the international literature and has been more dedicated to the studies we have carried out in a well defined population and area since the 1960's that is the resident population of the Health District of Ferrara, Northern Italy. The analysis of the findings of these studies we have carried out using a better and more proper definition of rural and urban environment and the different urban environment levels in the study area reveals some interesting findings that have induced us to modify our previous ideas about this issue providing some interesting clues that seem very important for the future studies on ALS and environmental factors which should consider the main actual change in the environment that is the urbanization and the progressive depopulation of the rural area.

Amyotrophic laterals sclerosis (ALS) is an idiopathic fatally progressive disorder of adult onset, which affects the lower motor neurons in the brainstem and the spinal cord, the upper motor neurons directly connected to them, and frequently the prefrontal motor neurons [1,5,7]. ALS is classified according to its clinical and epidemiologic features as sporadic accounting for 90-95% of the cases, familial accounting for 5-10% of the cases and Western Pacific (usually as ALS/PDC, ALS/parkinsonism-dementia complex), in three foci [1,5,7].

The cause of sporadic ALS is still unknown and probably is multifactorial. The clinical condition recognized as 'sporadic ALS' is probably the final common pathway for processes initiated in different ways [1,5,7]. Since the original description by JM Charcot at La Salpêtrière Hospital in Paris in 1874 [2,7], numerous clinicoepidemiological studies have been carried out, but the aetiology of ALS remains an enigma. The mean annual crude incidence rate of ALS is in the range from 0.4 to 2.6 per 100.000 population [1,3,4,5,7]. ALS is also known as "motor neuron disease", term proposed by Brain in 1962 [6]. Moreover, ALS is known in the United States as "Lou Gehrig's disease" after the famous New York baseball player who died in 1941, aged 37 years, of this disease . ALS represents most cases of adult motor neuron disease in all series (80-90%) and failure to separate ALS from other conditions may introduce a 10% error in incidence data and a larger one in prevalence data [5,7]. However, the term ALS and motor neuron disease are often used as synonyms [3,5,7]. This is due in part to the absence , until recently, of diagnostic criteria for ALS which is largely the major sporadic motor neuron disease of adults [5,7,8]. Our knowledge of the descriptive epidemiology of ALS is not complete. In many part of the world the occurrence of ALS has not been studied at all and even in Europe, Scandinavia and North America where a number of studies have been carried out, differences in the way they have been conducted constrains comparison of the results [5,7, 18]. Moreover, difficulties with clinical diagnosis, incomplete case ascertainment, and subreporting, all adversely affect epidemiological data on ALS. Furthermore, information about time trends of the disease mainly depends on routinely collected mortality data. While these data may be remarkably reliable for diseases with as high a case fatality as ALS, it is hard to be sure that changes in mortality over time only reflect true trends in the incidence of the disease rather than changes in diagnostic and certification practices. In the neuroepidemiological literature the prevailing view is that, with the notorious exception of the foci in the Western Pacific, the annual incidence of ALS is similar worldwide, about 1-2 cases/100.000 population [1,4,5,7,10,14,15,16]. However, it is unclear if annual incidence rates of ALS lower than 1 case/100.000 population represent true low disease incidence or limitations in case ascertainment [1,7, 10,14,15,16,18]. However, if one stops to consider the variety of conditions in which people live (and therefore the variety of the range of environmental factors to which they and their motor neurons are exposed) and the diversity of their genetic inheritance, the idea that any disease whatever its aetiology could be evenly distributed throughout the world seems scarcely credible. It would be important to find a true low incidence of ALS as it could involve the existence of some protective factor for ALS in that study area and population. However, in the field of neuropedemiology the prevailing view is that there are no robust environmental risk factors for ALS [1,14,15,16].The main purpose of studying the epidemiology of a disease is to identify the aetiology or the risk factors of that disease. By investigating the geographical distribution of a disease both on a global scale and within smaller areas and populations, changes in the frequency with which the disease occurs over time as well as the characteristics of the people who suffer from it, epidemiologists seek patterns that they hope will provide clues about its aetiology or about those factors that increase the risk of developing the disease. The task of summarizing the geographical distribution of ALS has been greatly facilitated by some reviews [3,4,5,7] which are recognized for their completeness and which summarized the result of most population-based studies of the disease. These reviews are especially valuable because the efforts made to ensure that they were comprehensive [3,4,5,7]. Only nine studies were judged by Chanchellor and Warlow in their review [3] to have achieved complete or nearly complete ascertainment of the cases of ALS in the study populations. The mean annual crude incidence rates from these nine studies range from 0.6 to 2.6 cases per 100.000 population, a 4.33 fold variation, a slight but not negligible variation. The wide variability of the published incidence rates could be interpreted based on the small study populations, thus leading to fluctuating rates and fairly wide confidence intervals (CIs) of the rates. In addition, different diagnostic criteria and methods of case ascertainment have been used, which may have led to over-or underascertainment of patients with ALS in the study populations. Large populations should be thus investigated using standard classification criteria to provide more stable incidence rates and make comparative findings available [10,11,12, 13,14]. Recently, fairly large ALS samples have been reported from European national or regional registries using the El Escorial diagnostic criteria [8]. These diagnostic criteria for ALS are unanimously recognized as a valuable instrument for the diagnosis and the categorization of the level of diagnostic accuracy. In these population studies based on prospective European national or regional registries [10-14] which are thought to achieve complete case ascertainment in the study population the estimated annual incidence rates range from 1.7 to 2.5 per 100.000 population, but this range of mean annual incidence rates of ALS from large population-based study based on prospective national or regional registries is still in the range of the incidence values of 0.6-2.6 cases per 100.000 population per year reported by Chanchellor and Warlow in their review [3]. However, it is still unclear if incidence rates lower than 1/100.000/year represent true low incidence of ALS or the consequence of limitations in case ascertainment in the studies [1,4,7]. Nonetheless, one of these studies which estimated an annual crude incidence of ALS lower than 1/100.000 deserves in our opinion special mention [17]. This was an incidence study on motor neuron disease over an 8-year period carried out in Mexico City amongst government workers and their dependants enrolled in a government-sponsored comprehensive health programme. This study estimated a very low annual crude incidence rate of 0.4 cases per 100.000 population which was not changed by the age-adjustment to the US population. The investigators restricted their attention to a subpopulation enrolled in a government health programme and the methods used to ascertain cases in the study population were at least as good as the methods used in studies which estimated incidence rates two- or threefold higher. Therefore, the range of annual incidence rates of ALS estimated in population studies which have achieved complete or nearly complete ascertainment of ALS cases is probably larger than that reported by Chanchellor and Warlow in their review [3], that is from 0.4 to 2.6 cases per 100.000 population per year which means a 6.5 fold variation, a not negligible variation. Such a variation seems to high to support the view that, with the exception of the foci in the Western Pacific, the incidence of ALS is similar worldwide. Most studies found that the incidence of ALS increases with age to a peak between 60 and 70 years of age and a decrease in the older decades [1,3,5,7].This pattern of the age-specific incidence of ALS might be interpreted to imply the role of a susceptible subpopulation but it could also imply the role of environmental factors in the aetiology of ALS [1,7]. In most studies, ALS predominated in males, with reported men to women ratios ranging from 1.2:1 to 2.6:1[1,3,4,5,7]. It has been suggested that this male preponderance could be the result of hormone influences or a confounder for putative risk factors such as occupational exposure, and physical activity [5,7].

Moreover, it has been suggested that the incidence rates of ALS in the industrialized countries have been increasing [1,4,7,9,15,16]. The incidence of ALS elsewhere may be rising although this could be due to improved ascertainment and demographic factors [1,4,7,15,16].The proponents of a hypothesys of increased incidence of ALS in recent years and in western countries have tended to imply the existence of some environmental cause for that increase may be associated with increased industrialization and urbanization [1,4,7,15,16]. Reports of ALS clusters are invariably fascinating to clinicians particularly in a fatal disease with unknown aetiology as ALS. In fact, the cluster could provide an important clue to the environmental causation of the disease. In general epidemiologists are more sceptical since they are aware of the fact that, in small populations, large fluctations in the frequency of uncommon diseases may occur by chanche [1,7,10,15,16].

Case-control studies comparing urban and rural populations in Italy found increased risk for ALS in rural populations which could indicate the role of exposure to agricultural chemicals [19,20]. However, two case-control studies, both carried out in large geographical areas showed no differences in the risk of ALS associated with urban and rural residence either at birth or at diagnosis [21,22]. Also in community-based surveys of motor neuron disease an increased risk in rural populations –reflected in two to seven times more cases found in the rural populations [24,25,27]. However, it is not surprising that the crude ALS incidence rates could be higher in rural populations compared to urban populations since the average age of rural populations at least in the western world is probably older than that in the urban populations because of the tendency for young people to move to cities so that the crude incidence rates should be age-adjusted. Moreover, neither of these studies was judged by Chanchellor and Warlow [3] to have used methods for the identification of the cases suitable for assuring complete case ascertainment. On the other hand, in another Italian population study the incidence of motor neuron disease was higher in the urban population than in the rural population [28].

Therefore, little information is available and there are no strong evidences that the occurrence of ALS could vary with the level of urbanization. A geographical analysis of all deaths from motor neuron disease in England and Wales during the period 1968-1978 using the usual place of residence as recorded on death certificates showed no significant differences between large towns (county boroughs), small towns (municipal boroughs and urban districts) and rural areas (rural districts) for mortality from motor neuron disease [18]. Moreover, a similar result was found in the USA when the age-adjusted mortality rates were mapped at county level [25]. No relation was apparent between mortality and percentage of the population living in urban areas. There was, however, a weak association between mortality and percentage of the population living on rural farms. Interestingly, in the field of the research about possible environmental factors involved in the pathogenesis of ALS,is that an apparent epidemic cluster of motor neuron disease has been reported in Sweden, occurring in men living in the county of Skaraborg in the years 1973-1984 [29,30]. Based initially on analysis of death certificates for Sweden in the period 1961-1990 [29], the study then actively searched all cases of motor neuron disease in this community and compared the results with those of the adjacent county of Värmaland [30]. During the period 1973-1984 an epidemic-like increase in cases among older men was found with an average incidence of 4/100.000 person-years (70 males identified), compared with an incidence of 2.8 for males and 1.6 for females in the entire period 1961-1990. These patients were found to have higher age at onset of motor neuron disease and more frequent than expected employment in agriculture [30]. Based on these results, an extensive case-control study of motor neuron disease was performed [31], showing that the highest odds ratio for the disease correspond to the combination of heredity (family history of neurodegenerative disease or thyroid disease), male sex, and exposure to solvents. On the basis of these studies, Gunnarsson suggested [32] that ALS may occur in a genetically susceptible population subset. These individuals may have defective detoxification leading to the neurodegenerative process precipitated by different agents (probably environmental agents) after an induction period of one to two decades [29,30]. Interestingly, a recent population study in Italy based on a prospective regional registry with virtually complete ascertainment of ALS cases in the study population [14] found a mean annual crude incidence rate of ALS of 2.5 per 100.000 and no differences in the incidence rates among the different provinces of this region were found. These findings and those of the other prospective studies based on a prospective regional or national registries [10-13] show that the ALS distribution in Western countries is remarkably similar and the perceived differences are likely to be mostly due to methodologic and demographic factors. These similar findings, despite the large differences in term of genetics, environment and socioeconomic background point to diffuse environmental or genetics factors rather than to specific exogenous toxins specifically acting in defined areas and populations (as could be the agricultural chemicals in a rural area) in the pathogenesis of ALS [14].

Therefore, there is little and conflicting information about the relationship between ALS and the level of urbanization. Interestingly, a protective effect for the Western pacific variant in Guam has been attributed to modernization [5,7,18]. The concept of modernization generally overlaps with a higher rate of urbanization. In fact, in the 2005 Revision of the UN World Urbanization Prospect report the 20th century has been described as "the rapid urbanization of the world's population "as the global proportion of the world urban population rose dramatically from 13% in 1900 to 49% in 2005 and this figure is likely to rise to 60% by 2030 since much less of the population is living in the rural area. Some surveys of classic sporadic ALS which compared contiguous urban and non urban populations have been carried out in Italy but one showed a significantly higher incidence rate of ALS in the non urban population [27] whereas another showed a significantly higher incidence rate in the urban population [28] and a third study disclosed no significant differences for that comparison [26]. In Italy during the time period covered by these studies the case ascertainment was probably lower in the non urban populations. More recently a prospective population-based study of ALS in the Republic of Ireland in the period 1995-1997 has found a mean annual crude incidence rate of 2.1/100.000 [12], this study has estimated a trend toward a higher incidence of ALS in the northeastern region of Ireland, an area which is isolated from the main population centres of Ireland and shows a lower population density than the rest of the country [12]. In this study there was not reason to suspect a different case ascertainment in the different regions of the study area.

When comparing different studies on ALS incidence a trend toward higher incidence rates in those areas with lower population density (Minnesota, Sweden and Finland) [36,37,38] compared to areas with higher population density such as Sardinia, Italy[39], Israel [40], Turin, Italy and Florence, Italy [28,34] has been pointed out [5]. This difference in ALS incidence between areas with lower population density (less urbanized) respect to areas with higher population density (more urbanized) was still present after the elderly was excluded [5]. However, these low density areas [36,37,38] are located at higher latitude than those areas with higher population density and share genetic traits of northern European populations so that some confounding by geographic latitude and race cannot be ruled out [5]. In fact, a latitude gradient from north to south for the incidence of ALS has been suggested by a comparison of the incidence rates at different latitudes in the USA and Europe [41] but the lower rates reported in the Mediterranean areas [39,40] could only reflect differences in case ascertainment and case definition. In fact, the latitude gradient hypothesis for ALS has been recently diminished by the findings of a population study based on a prospective regional registry in southern Italy which found an age-adjusted incidence of ALS of 1.7/100.000/year [13]. Therefore, the latitude gradient is likely not a confounder for the reported difference in ALS incidence between areas with low population density respect to areas with higher population density. Therefore, it is possible that the level of urbanization could be an environmental factor for ALS. Interestingly, Kondo in a national study from Japan which analyzed the mortality data from each of the 46 prefectures in the country restricting attention to women [42] concluded that a decreasing trend for motor neuron disease were more apparent in heavily industrialized areas which were also the most urbanized in the country. However, case-control studies of ALS in widespread populations have failed to show significant urban-rural differences either by place of residence at birth or at diagnosis [21,22]. However, the results may have been biased because of the use of prevalent cases and the inappropriate matching and choice of controls (spouses) for the survey intended or because death certificates identified the cases. Despite the ambiguities of the above mentioned results, it is possible that the level of urbanization could be an environmental factor involved in the pathogenesis of ALS and the low urbanization environment (that is the rural environment) could be a risk factor for ALS and a powerful confounder (e.g.,for farming, professional exposure to agricultural chemicals) not to be ignored [5].

The epidemiology of ALS has been intensively studied in the area and resident population of the Health District (HD) of Ferrara (a well-defined area and population of Northern Italy) with the aim of identify risk factors for the disease since 1960s by the staff of the Neurologic Clinic of the University of Ferrara. In a population-based case-control study [26] which involved the incident cases of ALS in the study population in the years 1964-1982 an association between ALS and rural residence (defined as residence in any type of centre with less than 5.000 inhabitants) at disease onset was found (OR= 1.8, 95% CI 1.1-3.1) as well as an association between ALS and agricultural occupations (OR=1.8, 95%CI 1.0-3.2). However , both the odds ratios were borderline and the definition of rural and urban residence could be misleading (less or more than 5.000 inhabitants) since it establishes a strict cut off for urban or rural residence without evaluation of the different levels of urbanization in the study area [18,59]. However, it is interesting that in the years 1964-1998 in two very small villages in the countryside in the HD of Ferrara with an average resident population during the years 1964-1998 of less than 1.000 inhabitants [44,59] the mean annual crude incidence rate of ALS was unexpectedly high, 7.65 /100.000 (95% CI of the rate 1.57-22.34) and 7.36/100.000/year (95% CI of the rate 1.5-21.5) respectively. In both these two very small villages the incidence rate has been based on few cases of ALS so that the estimated mean annual crude incidence rates show very large 95% confidence intervals. Therefore, these two very high incidence rates of ALS are probably only chance events due to the small numbers [1,7,10, 18]. As above mentioned the use of a strict definition of urban and rural residence is probably misleading and the different levels of urbanization in the study area in the study period should be considered [18].

A descriptive study in the HD of Ferrara in the years 1964-1975 estimated in the resident population of the HD of Ferrara a mean annual crude incidence rate of ALS of 1.07/100.000 population (95% CI of the rate 0.67-1.48) [26]. A following descriptive study estimated in the years 1976-1987 an average annual crude incidence rate of ALS of 1.61/100.000 (95% CI of the rate 1.10-2.27) [43]. A further incidence study in the HD of Ferrara found in the following years 1988-1998 an average annual crude incidence rate of ALS of 2.19/100.000 (95% CI of the rate 1.54-3.02) [44]. The relative increase of the crude incidence rate of ALS in the years 1964-1998 (the overall temporal span of our epidemiological descriptive studies on ALS in the HD of Ferrara) was 2.05 (95% CI of the relative increase 1.65-2.52; χ^2 test for trend = 6.4, $p<0.025$) which confirms a significant increase of the annual crude incidence of ALS in the study population and area over time in the years 1964-1998 [44]. An increase in both incidence and mortality of ALS during the last decades has been reported in several surveys in western countries [4,7, 9, 45-51]. This finding has been related to the growing population at risk for the disease because of the ageing of the population in western countries [1,7,51], but a true increase in the incidence of ALS over time cannot be excluded and this could imply the role of environmental factors in the pathogenesis of ALS [4,7]. However, the increase in the incidence of ALS reported in industrialized nations could also be the result of an improvement in ascertainment of ALS due to greater awareness of the disease [1, 7, 9, 15, 16]. However, the increase in the incidence of ALS over time has also been identified in areas and populations where no apparent changes in the health services have occurred over the corresponding years [36]. The reported increase in the incidence of ALS in a western area (Rochester, Minnesota) where high diagnostic capabilities for ALS were present throughout the study period 1925-1984 which did not reach the statistical significance [36] may be due to the small number of the incident cases of ALS.

The analysis of the findings in the HD of Ferrara in the years 1964-1998 [44] has shown that the increase in the annual crude incidence of ALS during the study period 1964-1998 was greater in women (relative increase of the incidence rate = 2.8, 95% CI of the relative increase 2.03-3.74; χ^2 test for trend = 6.70, $p<0.01$) than in men whose increase in the incidence of ALS was not significant (relative increase of the ALS incidence rate = 1.52, 95% CI of the relative increase 2.03-3.74; χ^2 test for trend =1.02, non significant) as well as the increase of the incidence of ALS was greater in elderly individuals (individuals of 70 years old and over, relative increase of the incidence rate = 3.78, 95% CI of the relative increase 2.27 -5.90; χ^2 test for trend = 4.49, $p<0.05$) than in adult individuals in the age group 30-69 years whose increase in the ALS incidence was not significant (relative increase of the incidence rate = 1.44, 95% CI of the relative increase 1.13 – 1.81; χ^2 test for trend = 1.45, non significant). Given the substantial population ageing occurred in the resident population of the HD of Ferrara during the study period 1964-1998 and the fact that the population ageing was more prominent in women than in men this increase in the annual crude incidence of ALS in the study population could be in short explained by the ageing of the population [44]. In fact, in the study population during the study period 1964-1998 the birth rate decreased from 16 to 6.2/1.000 inhabitants while the death rate increased from 9.7 to 13.5/1.000 inhabitants and as in most western populations the study population substantially aged over the study period (the individuals over 50 years of age increased from 23.8 % to 46.1% of the study population)[44]. Moreover, the old-age index (the ratio per 100 between the individuals in the age group 65 years and over and the individuals in the age group 0-14 years in the population) which is a synthetic indicator of the degree of population ageing has increased during the study period 1964-1998 from 61.3% to 286.1% (relative increase of the old-age index = 4.67, 95% CI of the relative increase 3.25-6.5) [44]. The average old-age index was 67.2% in 1964-1975, 109.1% in 1976-1987 and 256.9% in 1988-1998 a relative increase of 3.82 which confirms the above-mentioned population ageing in the study population. These demographic data support the hypothesis of the substantial role of the population ageing in the observed increase in the annual crude incidence of ALS in the years 1964-1998 in the HD of Ferrara. As above mentioned the observed increase in the crude incidence of ALS in the HD of Ferrara was greater in women than in men [44]. In fact, in men in the years 1964-1998 an increase in the crude incidence of ALS has been observed too, but unlike women it has not achieved statistical significance. An increase in ALS incidence and mortality among women during the last decades has been reported in other western areas [33, 45-51]. This interesting finding could be due to changes in lifestyle and behaviour among women in the last decades who could be more exposed to unknown environmental factors contributing to the pathogenesis of the disease than they had been in the past [4, 5,7]. Otherwise, if the increase in the annual incidence of ALS is simply related to population ageing, that is the so-colled Gompertzian effect or inter-disease competition, whereby better survival from high mortality conditions in old age (such as stroke and heart disease) would allow the expression of ALS in a greater number of surviving older individuals [7,33,51,53-57]. In fact, an increase in mortality has also been reported since 1950s for other degenerative neurological diseases of the elderly people such as Parkinson's disease, probably as a consequence of an increase in the susceptible aged population [7,58]. Therefore, if the increase in the incidence of ALS in western countries is the consequence of population ageing [7, 51], the greater increase in women than in men observed in the HD of Ferrara as in other western areas could simply be the consequence of a more prominent ageing in the subpopulation of women than in the subpopulation of men giving a greater increase in the incidence of ALS in women as compared with men [44,55]. The official statistics in the HDof Ferrara in the years 1964-1998 seem to support this hypothesis [44]. In fact the old-age index increased in the resident population of the HD of Ferrara in the years 1964-1998 more in women (from 73.8% to 361.9%; relative increase = 4.9, 95%CI of the relative increase 3.4-6.8) than in men (from 49.4% to 214.9%; relative increase = 4.3, 95%CI of the relative increase 2.9-5.9). The estimated mean values of the old-age index in the three following time periods in women were 81.61% in 1964-1975, 137.6% in 1976-1987, and 361.9% in 1988-1998 (relative increase = 4.43; 95%CI of the relative increase 1.01-12.7) whereas the estimated mean values of the old-age index in men in the three following time periods were 53.3% in 1964-1975, 82.0% in 1976-1987, and 196.4% in 1988-1998 in men (relative increase = 3.68; 95%CI of the relative increase 0.75-10.75). The age-adjustement procedure is used to remove the influence on the incidence estimate of the different age composition in the populations [60]. Therefore, given the observed substantial ageing over time in the years 1964-1998 of the resident population of the HD of Ferrara which has been, moreover, greater in women than men, the mean annual crude incidence rates of ALS estimated in the three following study periods 1964-1975, 1976 -1987 and 1988-1998 [26, 43, 44] have been age-adjusted by direct method using the European population as a standard [60,61]. Although the increase in the incidence of ALS was less evident with the age-adjusted rates than with the crude rates, an increase in the incidence of ALS over time in the whole study period 1964-1998 in the resident population of the HD of Ferrara has been confirmed with the age-adjusted rates which have been 0.99/100.000 (95% CI of the rate 0.62-1.49) in 1964-1975, 1.41/100.000 (95%CI of the rate 0.96-1.99) in 1976-1987, and 1.63/100.000 (95%CI of the rate 1.15-2.25) in 1988-1998 giving a relative increase of the age-adjusted incidence of ALS of 1.65 (95% CI of the relative increase 1.33-2.03; 99%CI of the relative increase 1.24-2.14). This increase over time in the years 1964-1998 of the age-adjusted incidence of ALS has been estimated in both sexes unlike the incidence crude rates of ALS which as above-mentioned have shown a significant increase in the incidence of ALS only in women (relative increase of the ALS crude incidence = 2.79; χ^2 test for trend = 6.70, p<0.01) whereas the increase was lower and did not achiewed statistical significance in men (relative increase of the crude incidence = 1.52; χ^2 test for trend = 1.09, not significant)[44]. In fact, in women the age-adjusted incidence of ALS increased from 0.77/100.000 (95% CI of the rate 0.35-1.46) in 1964-1975, to 1.12/100.000 (95% CI of the rate 0.62-1.88) in 1976-1987, and 1.58/100.000 (95% CI of the rate 0.98-2.42) in 1988-1998 giving a relative increase of 2.05 (95% CI of the relative increase 1.49-2.75; 99% CI of the relative increase 1.34-2.99) while in men the age-adjusted incidence of ALS increased from 1.22/100.000 (95% CI of the rate 0.65-2.08) in 1964-1975, to 1.7/100.000 (95% CI of the rate 1.04-2.76) in 1976-1987, and to 1.7/100.000 (95% CI of the rate 0.97-2.75) in 1988-1998 which gives a relative increase of 1.39 (95% CI of the relative increase 1.03-1.92; 99% CI of the relative increase 0.93-2.02). Although borderline and lower than the increase in the age-adjusted incidence of ALS observed in women the increase over time in the age-adjusted incidence of ALS has been found in men too on the contrary than with the crude incidence rates [44]. Therefore, in spite of the age-adjustement procedure which has been is used to remove the influence of the different age composition of the study population in the three study periods (since the substantial ageing of the resident population of the HD of Ferrara over time) the increase in the incidence of ALS has been confirmed and although borderline in men it has been estimated in both sexes. Therefore, although less apparent, an upward trend in the incidence of ALS over time in the population of the HD of Ferrara has been confirmed with the age-adjusted rates in the total population and in both sexes, although less evident in men. In our opinion these findings show that the increase in the incidence of ALS in the population of the HD of Ferrara cannot only be the consequence of the population ageing [1,7,9,15,16]. The increase in the incidence of ALS over time in the last decades has been observed in other industrialized nations and it has been considered as simply apparent [1,7,9,15,16] and apart from the population ageing and the decline in competing causes of mortality it has been considered as the possible consequence of other factors such as improved ascertainment due to the greater awareness of the disease, higher availability of techniques like electromyography as aids to diagnosis and better reporting of the disease [1,9,15,16]. So, the increasing incidence of ALS over time in western nations could be due to increased life expectancy, but other factors might also contribute such as better diagnosis since the El Escorial diagnostic criteria [8]. However, although it was not significant the incidence increase of ALS over time has been found in at least one western area (Rochester, Minnesota) where high diagnostic capabilities for ALS were present throughout [1, 36]. The studies on the incidence of ALS carried out in the area of the HD of Ferrara used the same study design, multiple sources of case material and the same case finding procedure and have been carried out on a well-defined population which is large enough to guarantee accurate case collection and minimise the risk of bias. Moreover, given the dramatic progression associated with ALS it does not seem unreasonable to assume that all cases would have at least one contact with one of the considered sources of case material [44]. Furthermore, the Italian National Health System provides medical care with free access to medical service so that there is no reason why the patient should not obtain medical care. Moreover, because the provision of benefits from the Italian National Health System and the Italian National Social Security is based on diagnosis and prescription by a neurologist practising in the area where the patient is resident, patients with ALS are really likely to be visited by a neurologist at least once during the course of the disease. The survey of non neurological sources minimised the possible loss of ALS cases in the three consecutive incidence studies [26,43,44] in the area of the HD of Ferrara. The third incidence study [44] has shown that only 2.2% of the incident ALS patients in the period 1964-1998 were cared elsewhere confirming that the neurological care is pratically self-contained within the community. So, it is unlikely that improved case ascertainment by itself could account for the observed ALS incidence increase over time in the HD of Ferrara [44]. Furthermore, the main difficulty of case ascertainment in ALS studies is in aged patients since they commonly live in nursing homes and may not be traced [10,63]. The difficulty of case ascertainment in aged patients with ALS is well-known [63]. Recent prospective studies based on regional registries which are thought to achieve complete ascertainment of the incident cases of ALS in the study populations [10,11,14] have found proportions of elderly patients of the incident cases of ALS in the study populations of 15% and 10% [10, 13]. In the population studies carried out in the HD of Ferrara the proportion of elderly patients was 22% of the identified incident cases of ALS [44]. This finding seems to confirm that the ascertainment of the incident cases of ALS in the population of the HD of Ferrara in the study period has been virtually complete. Furthermore, it is interesting that in the same area and population, repeated investigations on multiple sclerosis carried out by our own research group found stable incidence rates of MS over the years 1965-1993 [62]. On the other hand, better care of the aged might have led to a more accurate diagnosis of ALS in the elderly population so that the increase in ALS incidence could also be due to previous failure to diagnose the disease in the oldest age groups rather than poor ascertainment of the diagnosed patients and this could have mainly occurred in women [44,59]. The diagnosis of ALS could now be recognised more often than before because of better health services in general and neurological services in particular. However, better diagnosis of ALS in elderly women cannot be the sole factor since, although less apparent, an increase in the age-adjusted incidence of ALS has been observed in men too. Given the epidemiological findings in the area of the HD of Ferrara the role of environmental factors in the increase of the incidence of ALS cannot be excluded and should be considered. In fact, a true increase in the incidence of ALS in the population could be the consequence of unknown environmental factors [4, 7] that are not necessarily related to industrial toxins released in the environment [1,7, 15,16] as it has been debated in the past since the increase of the incidence of ALS observed in industrialized nations [1,4, 9,15, 16], but other factors related to the human activities and the profound changes in the environment due to the human activities could be involved. On the other hand, increasing emphasis is being placed on the genetic-environmental interaction model for ALS with age and time as modifier [7, 15,16, 64].

In a further survey carried out in the population of the HD of Ferrara in the years 1964-1998 with the aim of studying the possible relationship between the incidence of ALS in the population of the HD of Ferrara and the level of urbanization in the study area four different levels of urbanization have been considered using the official statistics data of the four municipalities in which the HD of Ferrara is subdivided [59]. The four considered urbanization levels have been from the most to the least urbanized level: towns > 100.000 inhabitants (that is the town of Ferrara), small towns 8.000-5.000 inhabitants, small towns 5.000-1.000 inhabitants, and "villages <1.000 inhabitants and scattered houses in the countryside". Using this stratification by these four levels of urbanization no differences have been found in ALS incidence among these four different levels of urbanization in the study area as well as no significant difference between the number of observed and expected cases of ALS in any of the four considered urbanization levels has been found [59]. Therefore, no correlation between the level of urbanization in the area of the HD of Ferrara and the incidence of ALS in the time period 1964-1998 has been found [59]. However, the analysis of the official statistics have shown that in the time period of the ALS incidence studies in the HD of Ferrara (1964-1998) in which an increase of the incidence of ALS in the population has been observed [26,43,44] a profound change in the residence pattern of the study population has occurred. In fact, a progressive decrease of the proportion of the population living in the proper rural area of the HD of Ferrara has occurred. In our opinion the proper rural level is only that defined as "scattered houses in the countryside" which means scattered isolated farmhouses in the countryside" (without any type of urban aggregation) . In our opinion only this definition identifies the proper rural level in the study area since any other type of definition reported in the official statistics (villages, small towns, towns) identify an urban aggregation more or less wide and therefore in any case an urban level (more or less urbanized) in the study area. The percentage of the study population living in this proper rural level of the HD of Ferrara "isolated scattered farms in the countryside" decreased from 21.4% in the years 1964-1975 to 17.6% in the years 1976-1987, and to 8.1% in the years 1988-1998 which gives an overall relative decrease of 62.2 %. At the same time the proportion of the study population which lives in an urbanized level of any type (villages, small towns, towns) in the HD of Ferrara increased from 78% of the study population in 1964-1975 to 83% in 1976-1987 and to 92% in 1988-1998. This data are the consequence of urbanization which is a well known demographic phenomenon, that is the progressive depopulation of the countryside and the tendency for people to move to cities and urban centres. Urbanization refers to a process in which an increasing proportion of an entire population lives in cities, urban centres and in the suburbs of cities. Urbanization is the increase in the population of cities in proportion to region's rural population. As above-mentioned the 2005 Revision of the UN World Urbanization Prospects report has described the 20th century as "the rapid urbanization of the World's population" since the global proportion of urban population rose from 13% in 1900 to 49% in 2005 and this figure is likely to rise to 60% by 2030 since much less of the population is living in the rural area. Therefore, the incidence of ALS in the population of the HD of Ferrara has progressively increased in the time period in which a progressive increase in the urbanization of the study area has occurred and in the time period in which a progressive decrease of the proportion of the study population living in the proper rural environment in the study area has occurred. So, it is possible to suppose a relationship between these events. In other words, it is possible to suppose that the increasing urbanization of the study area and the changes of the environment in which the study population lives could have been in some way involved in the increase of the incidence of ALS in the study population as well as the progressive decrease of the subpopulation living in the proper rural environment in the study area could have reduced the effect of some protective factor related to the rural environment in this proportion of the study population resulting in an increase of the incidence of ALS. Given these data some influence from the urbanized environment as well as from the proper rural environment in the population of the HD of Ferrara could be supposed. In the first descriptive epidemiological study of ALS carried out in the area of Ferrara [26] the average annual crude incidence rate of ALS in the years 1964-1982 in the urban population defined as people living in any type of centre with more then 5.000 inhabitants was 0.94/100.000 (95% CI of the rate 0.65-1.32) whereas the incidence of ALS in the rural population defined as people living in any type of centre with less than 5.000 inhabitants was 1.0/100.000 (95% CI of the rate 0.71-1.37) and the difference between the rates was not significant. As above mentioned, in our opinion this approach could be misleading since it establishes a strict cut-off between urban and rural population in the study area which could not correctly represent the real different levels of urbanization in the area [18]. The definition of the different levels of urbanization in the study area does not establish a strict cut off to distinguish between urban and rural population and makes possible to study the "strength" of any hypothetical association [1, 18, 59]. The residence pattern of the study population according the place of residence in the area of the HD of Ferrara in the years 1964-1998 has been rather different than that we have considered in our survey so that it is possible that the type of stratification by level of urbanization we have considered in our study has not been suitable [59]. The official statistics of the four municipalities in which the area of HD of Ferrara is subdivided show that in the years 1964-1998 an average of 13.290 individuals lived in isolated scattered farmhouses in the countryside without any type of urban aggregation which in our opinion identifies the proper rural level in the HD of Ferrara in 1964-1998; moreover, an average of 16.970 individuals lived in very small villages with less than 1.000 inhabitants which identifies a first mild level of urbanization in the study area (that is the lowest level of urban aggregation in the HD of Ferrara in 1964-1998); moreover, an average of 28.761 individuals lived in small towns > 1.000 inhabitants that identifies an higher level of urbanization respect to the previous and, finally, in the HD of Ferrara in the years 1964-1998 an average of 100.928 individuals (63.1% of he study population) lived in the town of Ferrara that identifies the highest level of urbanization in the study area. In the years 1964-1998 the distribution of the observed incident cases of ALS in the HD of Ferrara according to the level of urbanization of the place of residence at disease onset has been as follows: 3 cases of ALS in isolated scattered farmhouses in the countryside which gives an average annual crude incidence rate in this proper rural level of 0.64/100.000 (95% CI of the rate 0.13-1.85); moreover 11 incident cases of ALS have been observed in very small villages with less than 1.000 inhabitants which gives a higher average annual crude incidence rate of 1.85/100.000 (95% CI of the rate 0.92-3.31) and the difference between these two incidence rates is significant (difference between the rates = 1.21, 95%CI associated with the rate difference 1.15-1.27). Moreover, 18 incident cases of ALS have been observed in small towns with more than 1.000 inhabitants which gives an average annual crude incidence rate of 1.8/100.000 (95% CI of the rate 1.22-3.06) and also this incidence rate is significantly higher, although the 95%CIs of the two incidence rates overlap, than that observed in isolated scattered farmhouses in the countryside (difference between the rates = 1.16, 95%CI associated with the rate difference 1.13-1.17). Finally, 59 incident cases of ALS have been observed in the town of Ferrara giving an average annual crude incidence rate in this highly urbanized level in the study area of 1.7/100.000 (95% CI of the rate 1.27-2.15) and also this incidence rate is significantly higher than that estimated in isolated scattered farmhouses in the countryside (difference between the rates = 1.06, 95%CI associated with the rate difference 1.02-1.09). Furthermore, the crude annual incidence rate of ALS estimated in the true rural zone of the HD of Ferrara that is in isolated scattered farmhouses in the countryside (0.64/100.000, 95% CI of the rate 0.13-1.85) has been significantly lower than the incidence rate of ALS observed in the whole urban environment of the HD of Ferrara that is the three urbanized levels, from very small villages with less than 1.000 inhabitants to the town of Ferrara (on the whole, 88 incident cases of ALS, average pop. 146.659, annual crude incidence rate 1.71/100000, 95%CI of the rate 1.37-2.10; difference in rates 1.07, 95%CI associated with the rate difference 1.069-1.071). These data show that the incidence of ALS in the HD of Ferrara in the years 1964-1998 has not been homogeneous by urbanization level. Moreover, these data show that the proper rural environment could provide some protective effect respect to ALS and this unknown "protective effect" would be absent in any type of urban environment in the area. Furthermore, these data show that any type of urban environment in the study area presents an higher risk of ALS respect to the proper rural environment. As the tendency of young people to move to urban centres [18] the age composition of the populations living in these different urbanization levels of the study area in the years 1964-1998 has likely been rather different. Therefore, the mean annual crude incidence rates of ALS estimated in the four different urbanization levels we have considered in the HD of Ferrara in the years 1964-1998 have also been age-adjusted by direct method using the European population as a standard [60,61] (table 1). The annual age-adjusted incidence rates of ALS in 1964-1998 have been 0.45/100.000 (95%CI of the rate 0.09-1.16) in isolated scattered farmhouses in the countryside, 2.3/100.000 (95% CI of the rate 1.31-3.72) in small villages <1.000 inhabitants, 1.33/100.000 (0.78-2.14) in small towns >1.000 inhabitants and finally 1.5/100.000 (95%CI of the rate 1.20- 1.93) in the town of Ferrara. Therefore, the age-adjustment procedure has confirmed that the incidence of ALS in the HD of Ferrara in 1964-1998 has not been homogeneously distributed by urbanization level and that in the proper rural environment of the HD of Ferrara the incidence of ALS in the years 1964-1991 has been very low and it has been lower than the incidence of ALS in any of the urban levels of the study area. Considering together all the urban levels in the study area the age-adjusted incidence rate of ALS has been 1.52/100.000 (95% CI of the rate 1.22-1.87) higher than that observed in the proper rural environment level in the HD of Ferrara (0.45/100.000, 95%CI of the rate 0.09-1.16), although this difference has been only borderline. This, however, may be due to the very small number of incident cases of ALS observed in the proper rural level (only three cases of ALS in 35 years which gives a very low incidence rate of ALS with a large confidence interval). Therefore, the true conclusions about the findings in the HD of Ferrara using a better and proper definition of the less urbanized level that is the true rural level have been that in the true rural level the incidence of ALS has been very low and it has been lower than that of all the other identified urban levels and when this difference has not achieved the statistical significance with the age-adjusted incidence rates it has been likely due to the very small number of ALS incident cases observed in this proper rural level and the consequent very large 95% confidence interval of this very low incidence rate. So, the proper rural residence could be a true protective factor for ALS whereas the urban residence compared to the proper rural environment could be a risk factor for ALS.

Table 1. Age-adjusted annual incidence rate of ALS in the years 1964-1998 in the Health District of Ferrara, Northern Italy, according to the different levels of urbanization in the study area.

Isolated scattered farmhouses in the countryside	13.290	3 cases	0.45/100.000* (0.09-1.16)°
Very small villages with less than 1.000 inhabitants	16.970	11 cases	2.3/100.000 (1.31-3.72)
Small towns with more than 1.000 inhabitants	28.761	18 cases	1.33/100.000 (0,78-2.14)
Town with more than 100.000 inhabitants (the town of Ferrara)	100.928	59 cases	1.5/100.000 (1.20- 1.93)
All the urbanized levels of the study area considered as a whole§	146.659	88 cases	1.52/100.000 (1.22-1.87)

* The mean annual crude incidence rates of ALS estimated in the considered different levels of urbanization in the study area have been age-adjusted by direct method using the European population as a standard [60,61].

° 95% Confidence Interval.

§ very small villages with less than 1.000 inhabitants, small towns with more than 1.000 inhabitants and the town of Ferrara.

This analysis could imply different future important research approaches. In fact, although the strength of association is uncertain and there is not clear evidence of a "dose-response effect" (1,66) these findings suggest a tendency to a higher ALS incidence in the urban environment in the HD of Ferrara compared to the proper rural environment and this finding in a fatally progressive idiopathic disease such as ALS should be considered. However, association does not prove causation and it is basically to consider the rules of evidence such as the "chronological relationship" rule and the "biological plausibility" rule [1,66]. With regard to the "chronological relationship" rule there is an increased risk to disagree with this rule since the biological onset of ALS may precede the clinical onset of the disease by years or months [1,4,7,15,16]. With regard to the "biological plausibility" rule it is essential to define urbanization in term of environmental changes which possibly could be in some way harmful for motor neurons (e.g. higher concentration of human activities, for instance higher concentration of cars in the urbanized environment or higher concentration of heating plants or higher concentration of pollutants, etc.). In conclusion, the true conclusion from the epidemiological descriptive studies in the HD of Ferrara in a better and more proper definition of the less urbanized level that is the proper rural level in this study area is that in the true rural environment the incidence of ALS in the years 1964-1998 has been particularly low and it has been significantly lower than that observed in the urban environment levels. So, the proper rural environment "isolated scattered farmhouses in the countryside" could have been a protective factor for ALS whereas the urban environment compared to the proper rural environment could be a not negligible risk factor for ALS. This epidemiological findings could imply different future important research approaches.

References

[1] Armon, C. Motor Neuron Disease. In: Gorelick PB, Alter M (Eds): *Handbook of Neuroepidemiology*. New York: Dekker; 1994; 407-456.

[2] Charcot JM. Lectures on the diseases of the nervous system. Translated and edited by Sigerson G. Three volumes, second series. London: The New sydenham Society; 1881.

[3] Chanchellor, AM; Warlow, CP. Adult onset motor neuron disease:worldwide mortality, incidence and distribution since 1950. *Journal of Neurology, Neurosurgery, and Psychiatry*, 1992 55, 1106-1115.

[4] Worms, PM. The epidemiology of motor neuron disease:a review of recent studies. *Journal of the Neurological Sciences*, 2001 191, 3-9.

[5] De Pedro-Cuesta, J; Litvan, I. Epidemiology of motor neuron disease. In: Anderson DW, Schoenberg DG editors. *Neuroepidemiology. A tribute to Bruce Schoenberg*. Boca Raton: CRC Press;1991; 265-296.

[6] Brain, L. *Diseases of the Nervous System*. London: Oxford Press;1962.

[7] Roman, GC. Neuroepidemiology of amyotrophic lateral sclerosis: clues to aetiology and pathogenesis. *Journal of Neurology, Neurosurgery, and Psychiatry*, 1996 61, 131-137.

[8] Brooks, BR; Subcommittee on Motor Neuron Diseases/ Amyotrophic Lateral Sclerosis of the World Federation of Neurology research Group on Neuromuscular Diseases; the El Escorial "Clinical Limits of Amyotrophic Lateral Sclerosis" workshop contributors. El Escorial World Federation of Neurology criteria for the diagnosis of amyotrophic lateral sclerosis. *Journal of the Neurological Sciences*, 1994 124(suppl), 96-107.

[9] Juergens, SM; Kurland, LT; Okazaki, H; Mulder, DW. ALS in Rochester Minnesota, 1925-1997. *Neurology*, 1980 30, 463-470.

[10] Beghi, E; Millul, A; Micheli, A; Vitelli, E; Logroscino, G; for the Slalom Group. Incidence of ALS in Lombardy, Italy. *Neurology*, 2007 68,141-145.

[11] The Scottish Motor Neuron Disease Research Group. The Scottish Motor Neuron Disease Register: a prospective study of adult onset motor neuron disease in Scotland. Methodology, demography and clinical features of incident cases in 1989. *Journal of Neurology, Neurosurgery, and Psychiatry,* 1992 55, 536-541.

[12] Traynor, BJ; Codd, MB; Corr, B; Forde, C; Frost, E; Hardiman, O. Incidence and prevalence of ALS in Ireland,1995-1997. A population-based study. *Neurology,* 1999 52, 504-509.

[13] Logroscino, G; Beghi, E; Zoccolella, S; Palavano, R; Fraddosio, A; Simone, IL; Lamberti, P; Lepore, V; Serlenga, L; the SLAP registry. Incidence of amyotrophic lateral sclerosis in southern Italy: a population-based study. *Journal of Neurology, Neurosurgery, and Psychiatry,* 2005 76, 1094-1098.

[14] Piemonte and Valle d'Aosta Register for Amyotrophic Lateral Sclerosis (PARALS). Incidence of ALS in Italy. *Neurology,* 2001 56, 239-244.

[15] Armon, C. Environmental Risk Factors for Amyotrophic Lateral Sclerosis. *Neuroepidemiology*, 2001 20, 2-6.

[16] Armon, C. An evidence-based Medicine Approach to the Evaluation of the Role of Exogenous Risk Factors in Sporadic Amyotrophic Lateral Sclerosis. *Neuroepidemiology*, 2003 22, 217-228.

[17] Olivares, L; Esteban, ES; Alter, M. Mexican resistance to amyotrophic lateral sclerosis. *Archives of Neurology*, 1972 27, 397-402.

[18] Martyn, CN. Epidemiology. In: Williams AC, editor. *Motor Neuron Disease*. London: Chapman and Hall Medical; 1994; 393-426.

[19] Chiò, A; Maineri, P; Tribolo, A; Schiffer, D. Risk factors in motor neuron disease: a case control study. *Neuroepidemiology*, 1991 10, 174-184.

[20] Granieri, E; Rosati, G;, Paolino, E; Tola, R;, Ailello, I; Granieri, S. The risk of amyotrophic lateral sclerosis among laborers in Sardinia, Italy: a case-control study. *Journal of Neurology*, 1985 232 (suppl) 15.30.25, 241.

[21] Kurtzke, JF; Beebe, GW. Epidemiology of Amyotrophic lateral sclerosis 1. A case-control comparison based on ALS deaths. *Neurology*, 1980 30, 452-462.

[22] Kondo, K; Tsubaki, T. Case control studies of motor neuron disease. Association with mechanical injuries. *Archives of Neurology*, 1981 38, 220-226.

[23] Holloway, S; Emery, AE. The epidemiology of motor neuron disease in Scotland. *Muscle Nerve*, 1982 5(2), 131-133.

[24] Giagheddu, M; Puggioni,G; Masala, C; Biancu, F; Pirani, G; Piras, MB; Rachele, MG. Epidemiologic study of amyotrophic lateral sclerosis in Sardinia, Italy. *Acta Neurologica Scandinavica,* 1983 68(6), 394-404.

[25] Bharucha, NE; Schoenberg, BS; Raven, RH; Pickle, LW; Byar, DP; Mason, TJ. Geographic distribution of motor neuron disease and correlation with possible etiologic factors. *Neurology,* 1983 33, 911-915.

[26] Granieri, E; Carreras, M; Tola, R; Paolino, E; Tralli, G; Eleopra, R; Serra, G. Motor Neuron disease in the province of Ferrara, Italy, 1964-1982. *Neurology,* 1988 38, 1604-1608.

[27] Angelini, C; Armani, M; Bresolin, N. Incidence and risk factors of Motor Neuron Disease in the Venice and Padua districts of Italy,1972-1979. *Neuroepidemiology*, 1983 2, 236-240.

[28] Leone, M; Chio, A; Mortara, P; Rosso ,MG; Schiffer, D. Motor neuron disease in the province of Turin, Italy,1970-1980. *Acta Neurologica Scandinavica* , 1983 68, 316-327.

[29] Gunnarsson, LG; Lindberg, G; Söderfelt, B; Axelson, O. The mortality of motor neuron disease in Sweden 1961-1985. *Archives of Neurology*, 1990 47,42-46.

[30] Gunnarsson, LG; Lygner, PE; Veiga-Cabo, J; De-Pedro-Cuesta, J. An epidemic- like cluster of motor neuron disease in a Swedish county during the period 1973-1984. *Neuroepidemiology*, 1996 15, 142-152.

[31] Gunnarsson, LG; Bodin, G; Söderfelt, B; Axelson, O. A case-control study of motor neuron disease: its relation to heritability and occupational exposures, particularly to solvents. *British Journal of Internal Medicine*, 1992 49, 791-798.

[32] Gunnarsson, LG. Motor neuron disease and exposure to chemicals –aetiological suggestions from a case-control study. *Journal of the Neurological Sciences*, 1994 124 (suppl), 62-63.

[33] Neilson, S; Gunnarsson, LG; Robinson, I. Rising mortality from motor neuron disease in Sweden 1961-1990: the relative role of increased population life expectancy and environmental factors. *Acta Neurologica Scandinavica*, 1994 90, 150-159.

[34] Bracco, L; Antuono, P; Amaducci, L. Study of epidemiological and etiological factors of amyotrophic lateral sclerosis in the province of Florence, Italy. *Acta Neurologica Scandinavica*, 1979 60(2),112-124.
[35] Traynor, BJ; Codd, MB;, Corr, B; Forde, C; Frost, E; Hardiman, O. Incidence and prevalence of ALS in Ireland, 1995-1997. A population-based study. *Neurology*, 1999 52, 504-509.
[36] Yoshida, S; Mulder, DW; Kurland, LT; Chu, CP; Okazaki, H. Follow-up study of amyotrophic lateral sclerosis in Rochester, Minn., 1925 through 1984. *Neuroepidemiology*, 1986 5, 61-70.
[37] Forsgren, L; Almay, BG; Holmgren, G; Wall, S. Epidemiology of motor neuron disease in northern Sweden. *Acta Neurologica Scandinavica*, 1983 68(1), 20-29.
[38] Murros, K; Fogelholm, R. Amyotrophic lateral sclerosis in middle Finland: an epidemiological study. *Acta Neurologica Scandinavica*, 1983 67(1), 41-47.
[39] Rosati, G; Pinna, L; Granieri, E; Aiello, I; Tola, R; Agnetti, V; Pirisi, A; de Bastiani, P. Studies on epidemiological, clinical and etiological aspects of ALS disease in Sardinia, southern Italy. *Acta Neurologica Scandinavica*, 1977 55(3), 231-244.
[40] Kahana, E; Zilber, N. Changes in the incidence of amyotrophic lateral sclerosis in Israel. *Archives of Nurology*, 1984 41(2),157-160.
[41] Annegers, JF; Appel, S; Lee, JR; Perkins, P. Incidence and prevalence of amyotrophic lateral sclerosis in Harris County, Texas, 1985-1988. *Archives of Neurology*, 1991 48, 589-593.
[42] Kondo, K. Motor neuron disease: changing population patterns and clues for etiology. In: Schoenberg BD, editor. *Advances in Neurology*. New York: Raven Press; 1978; 509-543.
[43] Granieri, E; Rosati, G; Tola, MR; Paolino, E; Carreras, M; Monetti, VC; Aiello, I; Casetta, I; Govoni, V; Caniatti, L. Neurological diseases in the 1980s: descriptive neuroepidemiology in Ferrara province and Sardinia. *Epidemiologia e Prevenzione*, 1991 Sep-Dec13(48-49), 160-166.
[44] Govoni, V; Granieri, E; Capone, J; Manconi, M; Casetta, I. Incidence of Amyotrophic Lateral Sclerosis in the Local Health District of Ferrara, Italy, 1964-1998. *Neuroepidemiology*, 2003 22, 229-234.
[45] Buckley, J; Warlow, C; Smith, P; Hilton-Jones, D; Irvine, S; Tew, JR. Motor neuron disease in England and Wales, 1959-1979. *Journal of Neurology, Neurosurgery, and Psychiatry*, 1983 46,197-205.
[46] Durrelman, S; Alperovitch, A. Increasing trend of ALS in France and elsewhere: Are the changes real? *Neurology,* 1989 39, 768-773.
[47] Elian, M; Dean, G. The changing pattern of motor neuron disease and multiple sclerosis in England and Wales and the Republic of Ireland. *Neuroepidemiology,* 1992 11, 236-243.
[48] Lilienfeld, DE; Chan, E; Ehland, J; Godbold, J; Landrigan, PJ; Marsh, J; Pearl, DP. Rising mortality from motor neuron disease in the USA, 1962-1984. *Lancet*, 1989 Apr1 1(8640), 710-713.
[49] Maasilta, P; Jokelainen, M; Löytönen, M; Sabel, CE; Gatrell, AC. Mortality from amyotrophic lateral sclerosis in Finland, 1986-1995. *Acta Neurologica Scandinavica*, 2001 104, 232-235.

[50] Mueller Selljeseth, Y; Vollset, SE ; Tynes, OB. Increasing mortality from amyotrophic lateral sclerosis in Norway? *Neurology*, 2000 55, 1262-1266.

[51] Neilson, S; Robinson, I; Alperovitch, A. Rising amyotrophic lateral sclerosis mortality in France 1968-1990: increased life expectancy and inter- disease competition as an explanation. *Journal of Neurology*, 1994 241, 448-455.

[52] Kurtzke, JF. Risk factors in amyotrophic lateral sclerosis. In: Rowland LP editor. *Amyotrophic Lateral Sclerosis and other Motor Neuron Diseases Advances in Neurology*. Vol.56. New York: Raven Press; 1991; 245-270.

[53] Chiò, A; Magnani, C; Schiffer, D. Amyotrophic lateral sclerosis mortality in Italy, 1958 to 1987:A cross-sectional study. *Neurology*, 1993 43, 927-930.

[54] Neilson, S;Robinson, I; Hunter, M. Longitudinal Gompertzian analysis of ALS mortality in England and Wales, 1963-1989: estimates of susceptibility in the general population. *Mechanisms of ageing and development*, 1992 64, 201-216.

[55] Neilson, S; Robinson, I; Nymoen, EH. Longitudinal analysis of amyotrophic lateral sclerosis mortality in Norway 1966-1989: evidence for a susceptible subpopulation . *Journal of the Neurological Sciences*, 1994 122, 148-154.

[56] Neilson, S; Robinson, I; Kondo, K. A new analysis of mortality from motor neuron disease in Japan , 1950-1990: rise and fall in the postwar years. *Journal of the Neurological Sciences*, 1993 117, 46-53.

[57] Riggs, JE. Longitudinal Gompertzian analysis of amyotrophic lateral sclerosis mortality in the USA, 1977-1986: evidence for an inherently susceptible population subset. *Mechanisms of ageing and development*, 1990 55, 207-220.

[58] Romàn, GC; Zhang, ZX; Ellenberg, JH. The neuroepidemiology of Parkinsons's disease. In Ellenberg, JH; Koller, WC; Langston, JW editors. *Etiology of Parkinsons's disease*. New York: Dekker;1995; 203-243.

[59] Govoni, V; Granieri, E; Fallica, E; Casetta, I. Amyotrophic lateral sclerosis, rural environment and agricultural work in the Local Health District of Ferrara, Italy, in the years 1964-1998. *Journal of Neurology*, 2005 252, 1322-1327.

[60] Lilieenfeld, A.M. and Lilienfeld D.E. (1980) *Foundations of Epidemiology*. (second edition). New York: Oxford University Press.

[61] Waterhause, J; Muir, C; Shanmungaratham, K; Powel, J: Cancer incidence in five continents. *IARC scientific publications*, 1982 4: 673.

[62] Granieri, E; Malagù, S; Casetta, I; Tola, MR; Govoni, V; Paolino, E. Multiple Sclerosis in Italy. A reappraissal of incidence and prevalence in Ferrara. *Archives of Neurology*, 1996 53, 793-798.

[63] Forbes, RB; Colville, S; Swingler, RJ. The epidemiology of amyotrophic lateral sclerosis (ALS/MND) in people aged 80 or over. *Age aging*, 2004 33, 131-134.

[64] ALSA-Funded Workshop: Meeting Report. Environmental factors and genetic susceptibility in Amyotrophic lateral sclerosis May 29-31 ,2002, Keystone, Colo. http: //.alsa.org/research/workshops1.cfm.

[65] Ibrahim, ME. Rules of evidence. In Ibrahim, ME editor. *Epidemiology and Health Policy*. Aspen: Rockville MD; 1985.

In: Focus on Urbanization Trends
Editor: Laura M. De Smet

ISBN: 978-1-60456-109-8

Chapter 4

CHINA'S URBANIZATION: NEW TRENDS AND CHALLENGES OF CONCEPTUALIZATION

Li Zhang[*]
Department of Geography and Resource Management
The Chinese University of Hong Kong
New Territory, Hong Kong

Introduction

Economic reforms in socialist countries almost inevitably bring about significant changes in the spatial structure of the economy and urban development. China, once a socialist country undertaking systemic reforms on its centrally planned economy since the late 1970's, has experienced the shifting urbanization patterns with a scale and speed unparalleled its socialist period. Historical comparison of urbanization level shows that China is witnessing a revival of urban growth from its long-term stagnation of urbanization that occurred in the 1960's and 1970s. Starting from 172 million in 1978, by the end of 2003 China's urban population increased to 524 million, amounting to 47 per cent of the total population (NBS, 1983; 2005). This urbanization level is noticeably higher than it was at the beginning of the socialist period (11 per cent) and in the eve of systemic reforms (18 per cent). Very recently, urbanization has been officially conceived as a positive drive conducive to China's modernization. Chinese authorities have declared a clear desire to promote the country's urbanization as one of the overriding policy issues for the 11th Five-Year Plan and anticipated that, before too long, the number of urban population would exceed the rural population. While there is much to be applauded in this, China's current level of urbanization is still considerably lower than that of industrialized countries. The current level of urbanization in China is about the same as the world's average level in 2000 (United Nations Center for Human Development, 2002).

[*] E-mail address: lizhang@cuhk.edu.hk

In line with the rapidity of urbanization is the transformation of urbanization pattern. There are marked differences not only in the pace and the level but also in the driving forces of urbanization between the socialist and the post-socialist periods.[1] In the socialist period, China did not resemble the urbanization pattern found in many industrialized economies where there existed the strong correlation, statistically and functionally, between industrialization and urbanization. Rather, China witnessed industrialization with limited urbanization, largely as the result of government's anti-urbanization policies and restrictions on rural-to-urban migration (Kirkby, 1985; Chan, 1994; Zhang, 2004). The regulation of rural-to-urban migration was carried out by means of practices comparable to those for controlling international borders. By stark contrast, rural-to-urban migration and the development of city-regions are significant determinants of accelerating urbanization in post-socialist China. Rural-to-urban migration has accounted for over 70 per cent of urban population growth between 1978 and 2000 (Chan and Hu, 2003). The rising amount of rural-to-urban migration constitutes two categories of rural migrants classified along Chinese household registration (*hukou*) lines: those whose moves are through various government channels and with state-sanctioned residency rights at destination, and those whose moves are on a voluntary basis and without an official permission for the transfer of residential status. The latter migrants, who are statistically enumerated as part of the urban population in the national census but do not have a full right of abode in cities in substantive terms, is commonly known as 'floating population' or 'temporary population' under the existing socioeconomic institutions (Solinger, 1999).[2]

Viewed spatially, urbanization process in post-socialist China has manifested as not only the dynamic expansion of individual cities in population and land area but also the spatial clustering of the existing urban system (Fan, 1999; Ma, 2005). Urban expansion is seen as an increase in developed areas transforming from rural areas. Rapid urbanization has created extensive land-use changes and spatial expansion outward from the urban core, evidenced by a significant portion of farmland has been taken by industrial and urban developments (Smil, 1999; Ho and Lin, 2004). At the city scale, many cities are rapidly growing at their fringes, engulfing villages and farmland and transforming them into urban areas. At the regional scale, the official statistics disclosed uneven urbanization across the country. There is a trend of growing concentration of the urban population and economy in city-regions along the country's open and economically competitive Pacific coast. The geographical concentration

[1] The notion of 'post-socialist' in this paper is used in a broad sense to denote the phase of systemic transformation from centralization to decentralization, from command planning to marketization, and from isolation to globalization, but without changing the fundamentals of the socialist system and without clearly defined the ultimate destination. In China this phase corresponds to the period from 1978 to present. Many rules of the game have been rewritten on a trial-and-error basis during the systemic transformation. However, the notion of 'post-socialist' does not necessarily imply that China attempts to dismantle its underlying socialist ideology during the transformation from one economic system to another. After the inauguration of systemic reforms, the Chinese top leadership has always claimed that they are building a 'socialist market economy' with Chinese characteristics, which in essence means incorporating market forces within the limits set by the state.

[2] The terminology of 'floating population' or 'temporary population' often refers to those who are not supposed to legally stay in cities permanently, regardless of the duration of their residence (Ministry of Public Security, 1985). Floating population, in a statistical sense, include those traveling for the transient purpose, such as entrepreneurs traveling for business, tourists and people visiting relatives. In many occasions, floating population essentially refers to those working and residing away from their regular place of household registration (*hukou*) for a certain period, usually over 6 months (Goodkind and West, 2002; Liang and Ma, 2004). Politically and socially, floating population have restricted rights and entitlements outside their regular place of *hukou*.

of urban development in certain parts of China's vast space, particularly in the Pearl River Delta and the Yangtze River Delta, has been articulated as 'exo-urbanization' that emphasized the significant impact of globalization on the post-socialist urban transformation (Sit and Yang, 1997). Nationwide, the city system has also bee restructured through the reorganization of city space, known as "city administering county", "converting county to city", and annexing surrounding counties to the districts of a central city (Chung and Lam, 2004; Ma, 2005; Chung, 2007).

The unprecedented and intricate post-socialist urbanization process sets China apart not only from most countries but also from its early socialist state. As a result, many issues generated by the specific trajectory of China's urbanization remain elusive and present challenging opportunities for fruitful theoretical construction (Lin and Wei, 2002). By synthesizing and scrutinizing the latest official data, this paper examines recent changes in three spatial facets of China's urbanization, namely urbanization pace, rural-to-urban migration, and spatial reorganization of the urban system. On that basis, the paper identifies several issues which are important but remain under-discussion for theorization of China's post-socialist urbanization. The paper argues that to conceptualize Chinese post-socialist urbanization process requires, first, to explicate extant cutting-edge empirical studies and, then, to integrate their key propositions with a system-centered analytical framework.

The Rapidity of Urbanization

Urbanization process in post-socialist China has shown new pace and patterns that are quantitatively and qualitatively different from those under the Mao's socialist regime. Under the Mao's socialist regime, China's urbanization was stagnant. According to the Chinese statistical authority, at the beginning of the socialist period in 1949, 10.6% of China's population lived in the areas that were officially designated as urban (NBS, 1983, p.104). Though there was a short-lived episode of disordered or 'spurious urbanization' in the mid-1950's, the urban population increased by only 2.3 per cent per annum in the 1960's and 1970's.[3] 30 years later, just prior to the onset of socioeconomic transformation away from Mao's socialism, only 19 per cent of 975 million people were urbanites. With the initiation of systemic reforms, national development policy has become more and more pro-urban and China's urbanization has begun to accelerate. Since the early 1980s, cities have gradually and conditionally been open to the rural population. From a geographical point of view, at least three unaccustomed but seemingly bewildering urbanization trends have appeared under the new post-socialist political economy, which stand in shark contrast with those under socialism.

One salient trend is that the rate of urbanization has embarked on a fast track. This would not be disputable despite definitional inconsistencies and enumeration problems that have posed challenges for any attempts to more precisely estimate China's urbanization level over time (Chan and Xu, 1985; Ma and Cui, 1987; Zhang and Zhao, 1998; Chan and Hu, 2003; Zhou and Ma, 2005).[4] Table 1 presents an overview of urbanization trends in China since

[3] Zhou and Ma (2000) coined the notion of 'spurious urbanization' to refer to the superfluous growth of urban population during the period of Great Leap Forward.

[4] It notes that substantive efforts have been made by the Chinese statistics authority to rectify under-estimation or administrative inflation of urban population through several amendments of urban criteria (Zhou and Ma,

1978. As shown in the Table, it is no historical coincidence that the nation's level of urbanization has more than doubled over the past two decades. Whereas only 18 per cent of the population lived in urban areas in 1978, the proportion of urban dwellers rose to 42 per cent by 2004, with an increase of 24 percentage points. During that period, the urbanization rate increased by more than 1 per cent annually.

Table 1. China's urban population and urbanization level

Year	Urban population (ten thousands)	City population (ten thousands)	Town population (ten thousands)	% of urban population
	(1)	(2)	(3)	(4)
1978	17245			17.92
1980	19140			19.39
1985	25094	21228	3866	23.71
1997	39449	23905	15544	31.91
1998	41608	24959	16649	33.35
1999	43748	26018	17730	34.78
2000	45906	28583	17323	36.22
2001	48064	30401	17663	37.66
2002	50212	32924	17288	39.09
2003	52376	34197	18179	40.53
2004		35080		41.76

Notes:
1. The urban definition does not necessarily coincide from one period to another. City population refers to those living in city proper.
2. Town population is attained by subtracting city population from urban population.

Sources:
Columns (1) and (4): National Bureau of Statistics, China statistical yearbook, various years. Beijing: China Statistics Press.
Column (2): National Bureau of Statistics, China urban statistical yearbook, various years. Beijing: China Statistics Press.

Despite a substantial increase of urban population in the past 25 years, quite a number of researchers, based on a comparison of changes in China's urbanization level with its rate of economic restructuring, question that China's urbanization level has not risen at a rate as compatible as its level of income (e.g., Choi, 2000; Li, 2001; Guo, 2002; Chang, 2002; Chang and Brada, 2002; Zhang, 2004). This may seem surprising if reform-period urbanization should be primarily a function of economic factors, as commonly seen by many. China has been the world's fastest growing economy over much of the past two decades, with real GDP grew continuously at very high speed around 9 to 12 percent per year for more than 15 years (National Bureau of Statistics). Structurally, non-agricultural sectors have contributed about 85 per cent of total GDP. Not well in line with such economic trends, to date, China's urbanization level has been persistently lower than the world's average level (Table 2).

2005). However, as one could see from the discussion of this paper and elsewhere, the problems associated with the measurement of urban population originated from Chinese administrative and institutional settings. They are not merely academic questions by default, of concern only to accurate urban population counts. Rather they are momentous questions of concern to the competition for resources and the need for bureaucratic governance (Zhang and Zhao, 1998; Chung and Lam, 2004; Ma, 2005).

Table 2. International comparison of urbanization level

Year	World	More developed regions	Less developed regions	Less developed regions, excluding China	Least developed regions	China
1950	29.0	52.1	18.1	20.4	7.3	13.0
1955	30.8	55.2	19.7	22.2	8.3	14.2
1960	32.8	58.4	21.7	24.1	9.5	16.0
1965	34.7	61.5	23.7	26.3	11.1	17.6
1970	35.9	64.6	25.2	28.6	13.1	17.4
1975	37.2	66.9	26.9	31.0	14.9	17.4
1980	39.1	68.7	29.5	33.6	17.4	19.6
1985	41.0	69.9	32.3	35.9	19.2	23.0
1990	43.0	71.2	35.2	38.1	21.0	27.4
1995	44.8	72.3	37.7	39.9	22.8	31.4
2000	46.7	73.2	40.3	41.8	24.7	35.8
2005	48.7	74.1	42.9	43.6	26.7	40.4
2010	50.8	75.2	45.5	45.5	29.0	44.9
2015	52.9	76.4	48.0	47.6	31.6	49.2
2020	55.1	77.8	50.7	49.8	34.4	53.2
2025	57.5	79.3	53.4	52.3	37.5	56.9
2030	59.9	80.8	56.1	55.0	40.9	60.3

Notes:

1. The urban definition used for the computation of urbanization level varies country by country.
2. More developed regions: They comprise all regions of Europe plus Northern America, Australia/New Zealand and Japan.
3. Less developed regions: They comprise all regions of Africa, Asia (excluding Japan), Latin America and the Caribbean plus Melanesia, Micronesia and Polynesia (see definition of regions).
4. Least developed countries: Include 50 countries - Afghanistan, Angola, Bangladesh, Benin, Bhutan, Burkina Faso, Burundi, Cambodia, Cape Verde, Central African Republic, Chad, Comoros, Democratic Republic of the Congo, Djibouti, Equatorial Guinea, Eritrea, Ethiopia, Gambia, Guinea, Guinea-Bissau, Haiti, Kiribati, Lao People's Democratic Republic, Lesotho, Liberia, Madagascar, Malawi, Maldives, Mali, Mauritania, Mozambique, Myanmar, Nepal, Niger, Rwanda, Samoa, São Tomé and Príncipe, Senegal, Sierra Leone, Solomon Islands, Somalia, Sudan, Timor-Leste, Togo, Tuvalu, Uganda, United Republic of Tanzania, Vanuatu, Yemen and Zambia. These countries are also included in the less developed regions.

Source: Population Division of the Department of Economic and Social Affairs of the United Nations Secretariat, *World Population Prospects: The 2004 Revision* and *World Urbanization Prospects: The 2005 Revision*, http://esa.un.org/unup

It is also lower than that of many countries with a similar income level. Based on the UN projection (which would be conservative), China's urbanization level would not exceed the world's average level before 2030 (Table 2). To place China's urbanization-economy relationship in an international, comparative context, some observers have supposed that China's urbanization level was lower than would be expected for a country at China's level of development (Li, 2001; Chang, 2002). It seems that urbanization level in the post-socialist period has increased at a rapid pace relative to the socialist era, but, by the international standard, remains lower compared to the level of economic development.

Increase of Rural-to-Urban Migration

Rural-to-urban migration has become a permanent feature of Chinese society and significantly contributed to the rise of urbanization level in the post-socialist period. In China, the relative significance of the two components of urban growth (natural increase vis-à-vis migration) varied over different historical periods and it is extremely difficult to measure because of the lack of detailed and reliable data. Table 3 tabulates the results of quantitative estimates on the component of urban population growth between 1950 and 2000 by two of a fairly small number of available studies. The different estimates can be understandable from (1) the different definitions of 'urban' used; (2) the different kinds of data adduced and differences in their completeness; and (3) different methodologies brought to bear upon the estimations. Despite the existence of certain numerical discrepancies in two estimations, the points of commonality are that the growth of urban population in the socialist era was mainly the result of natural increase, particularly in the 1960s and 1970s (Table 3).

Table 3. Components of urban population growth, 1950-2000

Period	Natural increase (%)		Net rural-urban migration (%)	
	Chan and Hu	Liu at al	Chan and Hu	Liu et al.
1950-1957	40	53.3*	60	46.7*
1958-1960	18	20.7	82	79.3
1961-1965	NA	--	NA	--
1966-1977	69	63.41	31	36.59
1978-1982	22	27.31#	78	72.69#
1983-1990	26		74	
1991-2000	20		80	
1950-2000	33		67	

Notes:
* the figure refers to the period 1949-1957.
\# the figure refers to the period 1978-1999.
Sources: Chan and Hu (2003), p.62; Liu et al., p.19.

Such trend has been reversed in post-socialist China. One can observe a large scale of rural-to-urban migration that has dominated dynamic urban growth over the past 25 years or so.

Geographically, the main destination of rural-to-urban migration inflows is first confined in home-nearby townships which are the base of township and village enterprises (TVEs), and then is the coastal metropolises which are the main venues for foreign direct investment and where the urban economy is most dynamic. In the early stage of systemic transformation, TVEs had been actively promoted to play a leading role in rural industrialization and to be the key absorbent of surplus agricultural laborers who should 'leave the land (quit farming) but not the village and enter the factory but not the city'. In their golden age of development from the 1980s to mid-1990s, TEVs generated more than one-third of the country's GDP and provided over 90 per cent of rural non-agricultural employment, thus reducing the likelihood of migration to big cities (West, 1997).[5] TVEs, based on their viewpoint of operation costs and institutional security, were typically situated in respective villages and therefore were scattered over the countryside. With the role of TVEs, the rural economy was to industrialize, but generally not spatially agglomerate, leading to a more locally based, spatially dispersed, rather than a city-based development pattern on a national scale and 'urbanization from below' or 'in suit' urbanization (Ma and Fan, 1994; Zhu, 2000; Marton, 2000).[6]

The large presence of rural migrant workers in urban China began to occur in the late 1990s when the contribution of TVEs to rural non-agricultural employment decreased (Liu, et al., 2003). Many TVEs, inefficient with regard to resource utilization and having problems of technology know-how, experienced a sluggish growth. The situation has been exacerbated by increasing competitions from foreign-owned or city joint-venture firms. Consequently, surplus rural workers have been redirected to cities to seek employment opportunities. With the sluggish growth of TVEs and the penetration of FDI into coastal areas, large scale, long-distance, inter-provincial, city-bound moves have become a main flow of rural-to-urban migration, consisting of an important driving component of 'exo-urbanization' in certain parts of China (Sit and Yang, 1997).

Whether initiated by the government or by individuals, rural-to-urban migration in post-socialist China is primarily employment-driven. Therefore, the empirical data on the labor supply of urban employment can reveal, to certain extent, the important contribution of rural-to-urban migration to China's urbanization. Table 4 is the authoritative report of urban labor supply from rural areas. As the statistics show, rural-to-urban labor transfer, including those directly from rural areas and those institutionally changing their residential status, consisted of a significant part of new urban labor supply during the most time of the 1980s. The permanent rural-to-urban migration (column 3 of Table 4) served as a primary source. With a substantial increase in new urban labor supply since the 2000s (from 16.7 millions in 2000 to 33.5 millions in 2004), there were decreased shares of rural-urban labor transfer in new urban labor supply.[7] Nonetheless, the labor transfer increased in numbers with the unprecedented scale. In 2000 and thereafter, China's annual rural-to-urban migrant labor flow has been between 2 to 5 million people, and over half of them have permanently settled in urban areas. Also, since the mid-1990s, urban work-units recruited employees from rural areas almost two times more than those from urban areas to meet their demands for new labor (columns 5 and 6 of Table 4). The increasing number of rural-urban labor transfer implicitly reflects the active participation of rural migrants in China's urban labor market.

---

[5] However, some, using the 1 per cent sample of the 1990 China population census, argued that rural industrialization leading by TVEs did not significantly reduce inter-provincial urbanward migration during the period of 1985-1990 (Liang et al., 2002).

[6] Henderson's analysis (2002) on China's urban concentration came up with a similar point. He found that in the late 1990s the ration of large metro areas (with population over 3 million) to small metro areas (with population from 1 to 3 million) in China was not only much lower than worldwide ratio but also lower than that in 1960. The same analysis also discovered that in 2000 the spatial Gini coefficient of Chinese cities was lower than those of many countries, indicating a lower level of urban concentration prior to 2000.

[7] It is not clear about what factors accounted for such decrease, as the data in the Table are published with not fully informed of meanings of categorized sources. Such a situation presents obstacles to a better understanding of the quantitative contribution of rural-urban labor transfer to urbanization. Also, the accuracy and completeness of these data are uncertain. It is argued elsewhere that the official statistics failed to meaningfully capture a large amount of rural migrant workers who were not registered to relevant government agents for various reasons and therefore under-estimated the magnitude of rural-to-urban labor shift.

Table 4. Urban labor supply from rural areas

Year	New urban labor supply (ten thousands)			New employees of urban work-units (ten thousands)		
	Total	Transfer directly from rural areas	Conversion of the hukou status from rural to urban	Total	From rural areas	From urban areas
1988	844.3	159.9	422.6	532.9	68.9	154.7
1989	1055.9	92.3	100.3	691.1	113.1	192.0
1990	1122.2	74.2	67.8	710.5	109.8	229.0
1991	1147.7	82.2	65.6	776.5	124.4	247.5
1992	1208.33	98.06	135.78	742.1	137.0	223.1
1993	1244.66	97.48	126.53	1159.57	235.2	202.62
1994	1359.01	184.12	125.94	1113.21	231.05	180.764
1995	1412.59	161.89	138.84	1060.59	241.62	166.82
1996	1470.11	147.54	120.20	985.26	210.13	142.10
1997	1618.96	99.56	132.61	942.14	193.16	129.22
1998	1512.73	87.06	92.88	928.46	177.92	97.27
1999	1525.36	89.91	87.00	824.81	170.84	89.48
2000	1665.21	104.07	100.37	810.96	182.61	96.69
2001	1776.91	105.91	122.69	834.49	194.08	103.15
2002	2545.61	170.88	147.63	933.32	248.92	128.93
2003	2896.38	283.27	182.57	1034.21	309.78	151.69
2004	3350.69	341.42	166.52	1117.56	359.50	185.12

Note: New employees of urban work-units refer to those newcomers working in urban work-units of different ownership types.

Source: Department of Population and Employment Statistics, National Bureau of Statistics, Department of Planning and Finance, Ministry of Labor and Social Security (various years), China labor statistical yearbook. Beijing: China Statistics Press.

The floating population deluge, which is largely composed of rural migrants, most certainly represents an important force in China's post-socialist urbanization, though the precise magnitude is in dispute. The systemic transformation underway has significantly altered the very basis of state control over rural-to-urban migration. The urban employment right of rural migrants is partially legalized under the rubric of the 'temporary' urban residency policy.[8] Nationwide the floating population has become a major aspect of the Chinese society and economy. To the extent that rural-to-urban migration has become more and more spontaneous, the number of floating population is increasing year by year. The stock of China's floating population in urban areas was variously estimated, with a conservative guess around 80-120 million in the late 1990s and over 150 million recently (Solinger, 1999; Liu and Chan, 2001; Goodkind and West, 2002; Liang and Ma, 2004). Wang and Zuo (1999) estimated that one-fifth of the population in China's largest cities consisted of rural migrants. The large presence of floating population in urban China has created a new

[8] A nationwide 'temporary residence certificate' (TRC) system was introduced in urban areas in 1985. The key provision was that people of aged 16 and over who intended to stay in urban areas other than their place of *hukou* registration for more than three months were required to apply for a TRC. The TRC is valid for a certain period (usually one year) and is renewable. Under this system, urban employers can hire rural migrants with the TRC. For details, see Chan and Zhang (1999).

dimension of social and spatial stratification that is based on the right of urban residency, the issue that has been intensively discussed in the literature.

Substantial Reorganization of the Urban System

The fast pace of urbanization is accompanied by the spatial reorganization of the urban and regional system. Urban population has increasingly concentrated in cities. 48 per cent of the urban population growth during the period 1985-2003 has been absorbed by cities (Table 1). City population has now accounted for 65 per cent of the urban population. Although the absolute number of the population living in the myriad designated towns is still an important component of the urban population, it is increasing at a slower rate than the city population. It appears that the urbanization process in China is accelerated through the growth of city population.

Table 5. Number of cities, by population size, 1984-2003

Year	No. of designated cities						No. of designated towns
	Total	>2 millions	1--2 millions	0.5--1 million	0.2—0.5 million	<0.2 million	
1984	295	8	11	31	81	614	
1985	324	8	13	31	94	178	
1988	434	9	19	30	110	266	
1989	450	9	21	28	116	276	
1990	467	9	22	28	117	291	
1991	479	9	22	30	121	297	12,152
1992	479						
1993	570	10	22	36	160	342	
1994	622	10	22	41	175	374	
1995	640	10	22	43	192	373	
1996	666	11	23	44	195	393	
1997	668	12	22	47	205	382	
1998	668	13	24	49	205	377	
1999	667	13	24	49	216	365	19,244
2000	663	13	27	53	218	352	
2001	662	13	28	61	217	343	19,794
2002	660	33	138	279	171	39	
2003	660	33	141	274	172	40	20,113

Source: Socioeconomic Survey Team, National Bureau of Statistics (various years), Urban statistical yearbook of China. Beijing: China Statistics Press.

While cities have encompassed an increasing share of the population, the number of cities is also increasing. Table 5 tabulates the evolution of the hierarchy of Chinese city system by the population size. Overall, the numerical rise of cities has been much faster during the post-socialist period than during the socialist era. Between 1984 and 2004, 365 newly designated cities had been added to the hierarchical urban system, where at average 18.25 new cities were designated every year. In the hierarchy of the urban system, the number of cities with fewer than 200,000 inhabitants has decreased substantially in the past 20 years. The number of cities with 2 million habitants or more keeps on rising. At present, 45 per cent of China's cities were concentrated in the eastern region as opposed to 37 and 18 per cent in the central and western regions respectively. Correspondingly, the share of China's city population in the eastern region was 51 per cent, and 34 and 15 per cent in the central and western regions respectively. It is the eastern coast that is urbanizing most rapidly, with other regions lagging considerably behind.

During the new phase of urbanization, one can observe the extensive restructuring and reorganization of urban space at both regional and city scales. At a regional scale, global city-regions are in the making. A city-region morphologically refers to the spatial extent of economic interaction over a wide geographic region that is well beyond the jurisdictional administration of any single city (Scott, 2001). It contains dozens or even hundreds of jurisdictional subdivisions (municipalities, counties, and townships, etc.) lying in juxtaposition to one another, producing different and sometimes overlapping zones of influence at various spatial levels. From a geographic point of view, a city-region may be first developed from a *desakota* (an Indonesian term dubbed by Terry McGee for the convergence of village and city) and now represents a contiguous outgrowth of a set of formerly scattered urban centers and their immediate hinterlands (McGee, 1991; McGee and Robinson, 1995). Since the late 1990s, the emergence of Chinese global city-regions has become increasingly linked with the accumulation and circulation of global capital. Chinese global city-regions are characterized by massive inflows of FDI, active foreign trades, an increase in export-based manufacturing, a well-developed infrastructure of transport systems, and a dense concentration of migrant laborers.[9] These rapid and concentrated urban-economic growths have resulted in the juxtaposition of industrial, infrastructural and residential land uses occurring outside the principal city cores but between major urban and peri-urban areas, where, practically, cannot be easily classified as either urban or rural.

At the city scale, the most significant is a redrawing of municipal boundaries to bring rural areas into the urban ambit through the system of 'city administering county', the system of 'converting county into city', and the annexation of suburban counties (Chung and Lam, 2004; Ma, 2005). The system of 'city administering county' is to place one or more suburban counties under the jurisdiction of a neighboring administratively high-ranking city. The system of 'converting county into city' is that a whole county, which is basically rural by nature, is upgraded into city status without changing administrative boundaries. Table 6 provides information on the system of 'city administering county' and the system of 'converting county into city'. The change in the number of cities in the last two decades, as shown in Table 6, has been largely affected by the system of 'converting county into city'. From 1983 to 1999, some 380 counties became cities, these accounted for 90% of the total number of newly established cities during the period (Dai, 2000). The annexation of suburban counties refers to the change of suburban counties into city districts. A profound example is that Guangzhou, a leading city in the Pearl River Delta (PRD), annexed a number of its neighboring counties and county cities between 2000 to 2005, making the area of urban districts under the direct administration of the city government increase from 1,400 to 3,719

[9] As of 2004, three Chinese global city-regions, namely the Pearl River Delta, the Yangtze River Delta, and the Bohai Rim Region, accounted for 54 per cent of the national GDP, 79 per cent of the country's foreign trade, and 85 per cent of China's total inward FDI (NBS, 2005). Such regional concentration of economic outputs will not be changed much in near future.

square kilometers (Xu and Yeh, 2003; Yeung, 2005). Similar jurisdictional annexation can be observed in other major cities in the PRD (Table 7). While behind such reorganization there are several practical considerations and associated problems with regard to effective governance and resource allocation (Fan, 1999; Chung and Lam, 2004; Ma, 2005; Chung, 2007), spatial consequences of such reorganization have been the enlargement of many existing cities by jurisdictional change of city boundaries, and the arbitrary creation of many new cities that contain large tracts of rural land and substantial rural population. The 'city-not-city' and 'countryside-not-countryside' (*cheng bu cheng, xiang bu xiang*) landscape has emerged within the jurisdictional boundaries of many designated cities, bringing a strong rural dimension into China's post-socialist urbanization process.

To recapitulate, a substantive review here has demonstrated that post-socialist China has experienced some bewildering, distinct, and seemingly conflicting features of urbanization: a leap forward of urbanization concurrent with under-urbanized features, the massive and sustained flow of rural-to-urban migration concomitant with the bulk of migrants without urban citizenship, and the phenomenal reorganization of urban space and the proliferating addition of new cities with inclusion of ample rural attributes. Not surprisingly, many implications of such extensive and intricate urbanization will become major theoretical and policy issues deserving for further close examination and theoretical interrogation.

Table 6. Reorganization of the urban system

	Cities leading counties	
Year	**Number of cities leading counties**	**Number of counties under**
1949	3	3
1957	3	4
1958	29	120
1959	37	134
1960	48	234
1961	39	132
1962	35	127
1963	28	111
1965	25	79
1970	27	85
1977	42	133
1982	58	171
1983	126	534
1984	127	568
1985	146	658
1986	151	698
1988	169	711
1990	170	700
1993	184	741
1994	196	783
1995	196	771
1998	216	896
2000	247	1102
2001	252	1135
2002	262	1194

Table 6. Continued

	Cities changing from counties		
Year	Number of newly	Number of cities	As % of total newly
1978	2	0	0
1979	23	3	13
1980	7	2	29
1981	10	4	40
1982	12	5	42
1983	44	39	89
1984	12	11	92
1985	24	21	88
1986	29	26	90
1987	30	29	97
1988	53	48	91
1989	16	16	100
1990	17	15	88
1991	12	11	92
1992	38	37	97
1993	53	53	100
1994	53	53	100

Sources: Zhang and Zhao (1999); Chung (2007).

Table 7. Territorial expansion of selected cities resulting from annexation, PRD, 2000-2005

City	Original Area (km^2)	Newly Annexed Area (km^2)	Total Area (km^2)
Guangzhou	1,443.6	2,274.9	3,718.5
Foshan	157.3	3,713.8	3,867.5
Jiangmen	179.8	1,638.3	1,818.1
Huizhou	1,465.4	1,184.0	2,649.4
Zhuhai	723.6	674.8	1,653.0

Source: Adapted from Yeung (2005), p.8.

Conceptualization of China's Urbanization under Systemic Transformation

The previous sections have outlined the major tangible spatial manifestations of China's post-socialist urbanization. It appears that urbanization under systemic transformation has been taking on new patterns, both quantitatively and qualitatively, different from those of the socialist era, but somewhat like those found in other market economies. While some aspects of China's post-socialist urbanization have already received research attention, important theoretical questions remain: are the Chinese socialist urbanization and post-socialist one on contrasting different paths? How can we understand those apparently contradictory juxtapositions of urbanization changes at the conceptualization and theorization level? How can the study of China's post-socialist urbanization contribute to the understanding of diverse forms of global urban processes? The challenges to answer those questions should warrant the renewed interest in China's urbanization, a not new but still important topic.

In retrospect, there have already been serious attempts to theorize China's long-term stagnation in urbanization under socialism by modifying the conceptual model proposed by Ofer. The prevailing explanatory framework for the Chinese practice of urban containment has overwhelmingly focused on socialist industrialization strategy as an independent variable to determine a dependent variable – the urbanization pattern. In the early days of the People's Republic where the economic system was officially built in accordance with Marxist principles, the paradox between the Marxist assumption of the superiority of a communist system and its relatively uncongenial reality generated a political need for socialist 'forging-ahead industrialization'. In order for this to be successfully financed, however, the state created a centralized fiscal system directing state resources to expand the heavy industrial base on the one hand, and minimizing investment in consumption on the other. Accordingly, the city was construed as the site of production for non-consumer products rather than as a conglomeration of a heterogeneous population or as a place for organizing collective consumptions (Wu and Ma, 2005). Urbanization was viewed primarily as a cost incurred in the pursuit of 'forging-ahead' industrialization, since an increasing share of urban population would necessarily divert limited resources away from industrialization. It is because of the high investment priority given to industrialization at the expense of non-productive sectors that the rate of urban population growth had to be restrained by measures to contain urban growth, leading to the development pattern of industrialization without urbanization or under-urbanization. Conceptually, the resource impact of socialist development strategy was taken as the underlying logic for maximizing industrial output while simultaneously minimizing urban costs and China's under-urbanization in the socialist era was an inevitable and desirable concomitant of prodigious industrialization (Kirkby, 1985; Chan, 1994). Recently, extending the narrow confine of the widely-read perception that China's pre-reform urbanization process was largely a function of socialist industrialization, Zhang's work (2004) has shifted the focus of inquiry to the impact of broader systemic characteristics that shaped China's irregular relationship between development and urbanization.

Whereas China's limited urbanization under socialism has been relatively well conceptualized, the same cannot be said for reform-period urbanization. As the systemic reforms over the past two decades have changed almost all aspects of the centrally planned system of the Chinese economy, the historically specific socialist industrialization strategy is no long in place. Accordingly, a large body of literature generated by individual scholars in and outside China has been redirected to the changing elements of urbanization in post-socialist China. There are substantial empirical accounts on the evolution of China's ambiguous urban definitions and the impact on the estimations of urban population sizes and rural-to-urban migration (Chan and Xu, 1985; Cui and Ma, 1987; Zhang and Zhao, 1998; Chan and Hu, 2003; Zhou and Ma, 2005). Some studies draw attention to the process of dual-track urbanization, which combines a spontaneous, bottom-up, rural-based urbanization with a state-sponsored, top-down, and metropolitan-based urban growth (Guldin, 1997; Zhu, 1999; Shen, 2006). Some analyzes, though focusing on different regions and different aspects, are concentrated on endogenous and exogenous forces of urbanization and urban expansion (Sit and Yang, 1997; Lin, 1997; Marton, 2000; Zhang, 2002; Pannell, 2002; Lin and Ho, 2005). Some articulate the physical transformation of urban system and urban space through different scales and perspectives (Ma, 2005; Ma and Wu, 2005; Chung, 2007). Recently, Ma

(2002) and He et al. (2006) have comprehensively reviewed urban transformation in China and have proposed certain directions for pursuing further theoretical inquiries. Undoubtedly, those empirical and agenda-setting researches have provided insightful observations of reality and have enriched our understanding of China's road to urbanization.

Despite their profoundly different handling of a wide range of research questions and diverse strands of empirical findings, most of these endeavors, designed for revealing discursive practices, are fairly ad hoc, documentary, disaggregated and incomparable, and case-specific in nature, and have not come up with a convincing set of features and interpretations with a logically consistent theoretical framework. While the literature on China's post-socialist urbanization is on the upswing, they are empirically orientated, conducted largely in isolation from one another segmented by disciplinary boundaries (Ma, 2002). Few are puzzle-solving, coherent intellectual conjecture or speculation at a higher level of abstraction. Though rapidly changing economic and intellectual environment in China studies provides both pressing demands and rich opportunities for constructing theoretical discourses,[10] China's post-socialist urbanization has remained under-theorized, as most of the attention has been devoted to descriptive/empirical studies and has not been oriented towards synthesization and meaningful generalization.

As Chinese economy becomes more and more marketizing and globalizing, urbanization seems proceeding at the pace comparable to those experienced by market economies at similar stages. At the same time, Chinese urbanization patterns have resembled certain features found in other market economies, too. To the extent that rural-to-urban migration has become more a matter of personal expectations driven by the twin forces of rural push and urban pull, and migrants can rely on their own resources to survive in cities, many may believe that Chinese state has been sidelined in the urbanization process. The pattern of labor mobility has been linked to the significant but uneven impact of globalization and transnationalism in accordance with 'global shift' of capitalism (Dicken, 1998). The growth of global cities in China's urban system and the formation of city-regions has been viewed not so much a consequences of conscious design by the central state as a mixture of the decentralization of power and the globalization of capital (Logan, 2002; Shi and Hamnett, 2002). The situation of rural migrant labor in Chinese cities resembles those of 'undocumented workers' in Western cities (Solinger, 1999). The more market-driven urbanization has created the impression that China's urbanization has been no longer as unique as in its socialist era. The impression also seems to justify that China's urbanization pattern in the post-socialist period is apparently similar with that of other market economies.

In light of recent market changes, one is therefore forced to expect that urbanization theories based on market economies will become increasingly relevant to post-socialist China and will have a wider application for China's urbanization in the context of systemic transformation. It becomes a fashionable and intellectual tendency to apply western eloquent discourses/jargons and models to interpret different facets of urbanization unfolding in China. There is no shortage of cases in this respect. For example, the logic of the spatial restructuring of urban system and the reformatting of urban space has been dealt with by the notion of

[10] Several profound changes of economic and intellectual environment can be identified, namely, the dramatic growth of the China's status in the world, the renewed interest in area studies by mainstream scholars, the systematic release of statistical data by the Chinese authorities, and the emergence of a new generation of scholars with both up-to-date Western professional training and Chinese field experience. For elaboration, see Lin and Wei (2002), Yeung and Lin (2003).

'scale' (Ma, 2005; Chung, 2007). The operating mechanisms of urban spatial transformation, such as urban land development, have been interpreted by the 'growth machine' model and 'urban regime' analysis (Zhu J., 1999; Logan, 2002). The emergence of migrant communities within cities (known as *chengzhongcun*) in the course of urban physical expansion has been articulated through the 'power of place' (Ma and Xiang, 1998) and 'self help' (Zhang et al., 2003). Urban uncontrolled land development is considered as the unintentional and unexpected outcome of 'institutional fixes' (Lin and Ho, 2005). These 'theoretical' inquiries reflect attempts to contextualize the Western concepts/theories in the Chinese diverse and complex urban changes. But at present no mature theory has been developed that can adequately account for the complexity of China's urbanization and convincingly explain seemingly conflicting reality outlined earlier. We even understand little the new rules of the game that have governed China's urbanization in transition.

Concluding Remarks: Unsettled Questions

Despite the fact that Chinese urbanization in the post-socialist period exhibits many similarities to corresponding phenomena in other market economies undergoing similar processes, at the macro level it also manifests certain unique Chinese features. The idiosyncrasies of China's post-socialist urbanization can be argued through several murky and unsettled questions.

The first is a perplexing picture of China's urbanization-economy relationship. Two contrasting views exist. Viewing the remarkable growth of Chinese economy and the concurrent acceleration of rural-to-urban migration, many believe that Chinese urbanization level is now not out of line with its economic development. By comparing and contrasting the Chinese case with the world's trend, some argue that, by the international 'norm', Chinese economic growth produced less than a commensurate urbanization level and urbanization rate is slower-than-expected even in the post-socialist period (Li, 2001; Guo, 2002; Chang, 2002; Chang and Brada, 2002). By quantifying a causal link between urbanization and economic development over time as well as across countries, Zhang (2004) found that a smaller contribution of GDP growth to a rise of urbanization level in China, relative to other countries either with the similar level of development or with the similar systemic features. Even with the category of socialist or transitional economies that are perceived as generally under-urbanized (Sjoberg, 1992; Szelenyi, 1996), surprisingly, China appeared much under-urbanized. Those conflicting arguments need to be further interrogated and validated. Quite aside from the verification of specific arguments made by the above studies, the existence of seemingly competing views with regard to China's urbanization-economy relationship raises an interesting question whether China maintains an idiosyncratic pattern of urbanization in the post-socialist period.

Another is a paradoxical reality that rural migrants can, are often welcome to, work in cities under certain restrictions but are deprived of their urban citizenship. The rural-urban divide has been blurred and Chinese society has become more fluid since the systemic reforms launched in the late 1970s. It is easily observable that rural laborers are no longer entirely prohibited from working in the city. Associated with this development is the rise of rural-to-urban migration outlined earlier. While reckoning with an irreversible trend toward further relaxation of official control over urban employment, city governments have always sought to maintain their grip on granting the permanent right of urban abode to rural migrants for various concerns. The urban citizenship of rural migrant workers is tenuous under the existing *hukou* system. The result is the large presence of rural migrants working in cities without officially lodged residency status. It has been widely reported that rural migrant workers have, economically, socially and spatially, been marginalized in urban China (Zhang et al., 2007). Much has been said on the peripheral status of rural migrant workers in cities. There has been also rising concern about their citizen rights expressed in both media reports and Chinese academia recently. Such concern stems from the perception that unequal treatment of rural migrant workers and the denial of their permanent right of urban abode are both inhumane and unjust. Despite much discussion and debate, and even talk of fundamental reform, the authorities are reluctant to formally recognize rural migrants as a part of the legitimate urban population with the same entitlements and obligations as those with the permanent right of abode in cities. Meanwhile, few studies explore the question, on a theoretical and even an empirical ground, as to why, despite the possibility of systemic reform, rural migrant laborers, who are no longer strictly barred from seeking jobs in cities, are not granted the same urban residency rights as those with urban *hukou*.

The last but not the least is the equivocal roles/powers of state and the multiple market forces in deciding the nature and pattern of development in general and urbanization in particular. Existing studies recognize the strong power of market on the one hand, and emphasize the predominant role of the state on the other. Some conceive that under the context of globalization and systemic reforms, market forces are more and more powerful and the omnipresent state is now being substituted by complex interactions among different actors of the economy (central and local governments, domestic enterprises, and foreign investors). In this regard, some opine that extant researches tend to emphasize the unusual aspects of Chinese cities and overlook the fact that the transformation of Chinese cities is part of wide global capitalist changes (He et al., 2006, p.446). By emphasizing the fundamental nature of systemic reforms, some argue that, despite the presence of other forces in China's socioeconomic changes, such as globalization and market with which the state must interact, the state remains the director and the leading actor in the unfolding drama of economic transformation (Ma, 2002, p.1562). In all spheres of reforms, the state is not only the ultimate policymaker and the manipulator of institutions but it also the controller which evaluates and approves major project applications and oversees their implementation (Ma, 2002). Although state's composition and functions are no longer the same as they used to be, the socialist state remains in command. The importance of the Chinese party-state as an invisible yet powerful hand responsible for the formation and reformation of China's restless landscapes has been recognized by almost all Chinese geographers. One team of researchers has even said 'bringing the state back in' (cf. Lin and Wei, 2002, p.1538). If the articulations of both sides may represent at least a plausible story of the complex reality, what left unclear is the specific role the state plays in the processes of development and urbanization. Furthermore, the legitimization of state's paramount role and power has long been taken for granted that Chinese state is unequivocally more powerful in all development issues. On the intellectual front, few extant researches have offered understanding how the paramount role and 'undivided power' of the state have been legitimized and materialized under the specific institutional setting of China's economic system.

The unsettled questions addressed here present challenges for the constructive theorization of China's post-socialist urbanization. The spatial transformation of post-socialist urbanization reflects the relational interplay and the uneasy blending of the legacy of pre-reform systemic features and external and internal market forces. There are two implications. First, the curious mixture of socialist and market elements in the Chinese system is such that only selected features of Western models are applicable. The causal factors underlying the patterns of urbanization in post-socialist China are rather unlike those to be found in other market economies. Second, as urbanization process is fundamentally shaped by the nature of political and economic system, it becomes logical to take system-centered political economy perspectives to investigate urbanization in the post-socialist period.

References

Chan K W and Hu Y, 2003, "Urbanization in China in the 1990s: new definition, different series, and revised trends" *The China Review* **3** 49-71.

Chan K W and Xu X, 1985, "Urban population growth and urbanization in China since 1949: reconstructing a baseline" *China Quarterly* **104** 583-613.

Chan K W and Zhang L, 1999, "The hukou system and rural-urban migration in China: processes and changes" *China Quarterly* **160** 18-55.

Chan K W, 1994 *Cities with invisible walls: reinterpreting urbanization in post-1949 China* (Oxford University Press, Hong Kong).

Chang G H and Brada J C, 2002, *China's urbanization lag during the period of reform: a paradox* (Working Paper, Department of Economics, The University of Toledo) *http://www.utoledo.edu/~gchang/publication/lagurban4.pdf*

Chang G H, 2002, *Test of the urbanization lag of China and its cost in employment* (Working Paper, Department of Economics, The University of Toledo).

Choi S, 2000, *Agenda for China's urbanization policy: economic mobility and integration* (Paper presented at the workshop on "China's urbanization strategy – opportunities, issues and policy options", May 8-10, 2000, Beijing).

Chung H, 2007, "The change in China's state governance and its effects upon urban scale" *Environment and Planning A* **39** 789-809.

Chung J H and Lam T, 2004, "China's "city system" in flux: explaining post-Mao administrative changes" *China Quarterly* **180** 945-964.

Dai J, 2000 *Zhongguo shizhi (The city system of China)* (Zhongguo ditu chubanshe, Beijing).

Dicken P, 1998 *Global shift: transforming the world economy (third edition)* (Paul Chapman Publishing Ltd).

Fan C, 1999, "The vertical and horizontal expansions of China's city system" *Urban Geography* **20** 493-515.

Goodkind D and West L A, 2002, "China's floating population: definitions, data and recent findings" *Urban Studies* **39** 2237-2250.

Guldin G E, Ed 1997 *Farewell to peasant China: rural urbanization and social change in the late twentieth century* (M.E. Sharpe).

Guo K, 2002, "Gongyehua yu chengshihua de jingjixue fengxi (Economic analysis of the relationship between industrialization and urbanization)" *China Social Science* **2** 44-55.

He S, Li Z, and Wu F, 2006, "Transformation of the Chinese city, 1995-2005: geographical perspectives and geographers' contributions" *China Information* xx 429-456.

Henderson V, 2002, *Urbanization in China* (Brown University, November Mimeo).

Ho Samuel P S and Lin George C S, 2004, "Converting land to nonagricultural use in China's coastal provinces: evidences from Jiangsu" *Modern China* **30** 81-112.
Kirkby R J R, 1985, *Urbanization in China: town and country in a developing economy 1949-2000 AD* (Croom Helm, London).
Li W, 2001, "*Chengshihua zhihou de jingji houguo fenxi (*Analysis of the impact of urbanization lag on economy)" *China Social Science* **4** 64-75.
Liang Z and Ma Z, 2004, "China's floating population: new evidence from the 2000 census" *Population and Development Review* **30** 467-488.
Liang Z, Chen Y P, and Gu Y, 2002, "Rural industrialization and internal migration in China" *Urban Studies* **39** 2175-2187.
Lin George C S and Ho Samuel P S, 2005, "The state, land system, and land development processes in contemporary China" *Annals of Association of American Geographers* **95** 411-436.
Lin George C S and Wei Dennis Y H, 2002, "China's restless urban landscapes 1: new challenges for theoretical reconstruction" *Environment and Planning A* **34** 1535-1544.
Lin George C S, 1997 *Red capitalism in south China: growth and development of the Pearl River Delta* (UBC Press, Vancouver).
Liu S, Li X, and Zhang M, 2003 *Scenario analysis on urbanization and rural-urban migration in China* (interim report for International Institute for Applied Systems Analysis, Austria (IR-03-036)).
Liu T and Chan K W, 2001, "National statistics on internal migration in China: comparability problems" *China Information* **15** 75-113.
Logan J R, (ed.) 2002 *The new Chinese city: globalization and market reform* (Blackwell, Oxford).
Ma Laurence J C and Cui G, 1987, "Administrative changes and urban population in China" *Annals of the Association of American Geographers* **77** 373-395.
Ma Laurence J C and Fan M, 1994, "Urbanization from below: the growth of towns in Jiangsu, China" *Urban Studies* **31** 1625-1645.
Ma Laurence J C and Wu F, Eds 2005 *Restructuring the Chinese city: changing society, economy and space* (Routledge, London).
Ma Laurence J C and Xiang B, 1998, "Native place, migration and the emergence of peasant enclaves in Beijing" *China Quarterly* **155** 546-581.
Ma Laurence J C, 2002, "Urban transformation in China, 1949-2000: a review and research agenda" *Environment and Planning A* **34** 1545-1569.
Ma Laurence J C, 2005, "Urban administrative restructuring, changing scale relations and local economic development in China" *Political Geography* **24** 477-497.
Marton A M, 2000 *China's spatial economic development: regional transformation in the lower Yangzi Delta* (Routledge, London).
McGee T and Robinson I, Eds 1995 *The new southeast Asia: managing the mega-urban regions* (UBC Press, Vancouver).
McGee T, 1991, The emergence of desakota regions in Asia: expanding a hypothesis, in *The extended metropolis: settlement transition in Asia* Eds Ginsberg N, Koppel B, and McGee T (University of Hawaii Press, Honolulu) pp3-25.
Ministry of Public Security, 1985, *Provisional regulations on the administration of temporary urban population* (issued on July 13, 1985).

National Bureau of Statistics (NBS), 1983 *China statistical yearbook 1983* (China Statistics Publishing House, Beijing).

National Bureau of Statistics (NBS), 2005 *China statistical yearbook 2005* (China Statistics Press, Beijing).

Pannell, C W, 2002, "China's continuing urban transition" *Environment and Planning A* **34** 1571-1589.

Scott A J, 2001, "Globalization and the rise of city-regions" *European Planning Studies* **9** 813-826.

Shen J F, 2006, "Understanding dual-track urbanization in post-reform China: conceptual framework and empirical analysis" *Population, Space and Place* 12 497-516.

Shi Y and Hamnett C, 2002, "The potential and prospect for global cities in China: in the context of the world system" *Geoforum* **33** 121-135.

Sit Victor F S and Yang C, 1997, "Foreign-investment-induced exo-urbanization in the Pearl River Delta, China" *Urban Studies* **34** 647-677.

Sjoberg O, 1992, "Underurbanization and the zero urban growth hypothesis: diverted migration in Albania" *Geografiska Annaler B* **74B** 3-19.

Smil V, 1999, "China's agricultural land" *China Quarterly* **158** 414-429.

Solinger D, 1999 *Contesting citizenship in urban China: peasant migrants, the state, and the logic of the market* (University of California Press, Berkeley).

Szelenyi I, 1996, "Cities under socialism – and after", in *Cities after socialism: urban and regional change and conflict in post-socialist societies* Eds. Andrusz G, Harloe M, and Szelenyi I (Blackwell Publishers) pp 286-317.

United Nations Center for Human Development, 2002 *The state of the world cities report 2001* (United Nations, New York).

Wang F and Zuo X, 1999, "Inside China's cities: institutional barriers and opportunities for urban migrants" *American Economic Review, Papers and Proceedings* **89** 276-280.

West L A, 1997, "Shifting boundaries" *The China Business Review* September/October Issue 15-20.

Wu F and Ma Laurence J C, 2005, "The Chinese city in transition: towards theorizing China's urban restructuring", in *Restructuring the Chinese city: changing society, economy and space* Eds. Ma Laurence J C and Wu F (Routledge) pp 260-279.

Xu J and Yeh A, 2003, "City profile: Guangzhou" *Cities* **20** 361-374.

Yeung H W and Lin George C S, 2003, "Theorizing economic geographies of Asia" *Economic Geography* **79** 107-128.

Yeung Y M, 2005, *The Pearl River Delta mega urban-region: internal dynamics and external linkages* (Occasional Paper no.12, Shanghai-Hong Kong Development Institute, the Chinese University of Hong Kong).

Zhang Heather X, Wu B, and Sanders R, (eds.) 2007 *Marginalisation in China: perspectives on transition and globalisatio.* (Ashgate).

Zhang Kevin H, 2002, "What explains China's rising urbanization in the reform era?" *Urban Studies* **39** 2301-2315.

Zhang L and Zhao Simon X B, 1998, "Re-examining China's "urban" concept and the level of urbanization" *China Quarterly* **154** 330-381.

Zhang L, 2004 *China's limited urbanization: under socialism and beyond* (Nova Science Publishers, Inc., New York).

Zhang L, Zhao Simon X B, and Tian J P, 2003, "Self-help in housing and chengzhongcun in China's urbanization" *International Journal of Urban and Regional Research* **27** 912-937.

Zhou Y X and Ma Laurence J C, 2000, "Economic restructuring and suburbanization in China" *Urban Geography* **21** 205-236.

Zhou Y X and Ma Laurence J C, 2005, "China's urban population statistics: a critical evaluation" *Eurasian Geography and Economics* **46** 272-289.

Zhu J, 1999, "Local growth coalition: the context and implications of China's gradualist urban land reforms" *International Journal of Urban and Regional Research* **23** 534-552.

Zhu Y, 2000 *New paths to urbanization in China: seeking more balanced patterns* (Nova Science Publishers, Inc., New York).

In: Focus on Urbanization Trends
Editor: Laura M. De Smet
ISBN: 978-1-60456-109-8

Chapter 5

BIODIVERSITY AND ECOLOGY OF HYDROBIONTS IN NATURAL AND ANTHROPOGENIC LANDSCAPES

I.M. Khokhutkin, L.A. Kovalchuk and L.V. Chernaya
Institute of Plant and Animal Ecology,
Ural Division of Russian Academy of Sciences

Abstract

Usually only terrestrial habitats are considered in the literature concerning investigations of the urban fauna ecology, while freshwater ecosystems are omitted. Leaches, benthic invertebrates, play a very important part in water ecosystems of various types. They are an important link in trophic chains: predators and blood-suckers for various groups of hydrobionts they are a prey for many fish species, desmans, otters, minks, etc. Some predatory leach species play a "sanitary" role in reservoirs, as they prey on carrion. Besides, they are additional and intermediate hosts for many parasites. The aim of the work is to investigate the hirudofauna of reservoirs of various types (rivers, lakes, ponds, water reservoirs) in the South and Middle Urals. We compared the species composition and ecologo-physiological characteristics of leaches from 28 reservoirs located in urbanized and natural areas. We revealed the effect of anthropic factors on the qualitative and quantitative compositions of the leach fauna in the urban reservoirs.

Increased concentrations of pollutants (heavy metals) are marked in animal has tissues against the background of a decreased number of species and their abundance. This resulted from changes in the habitat and feeding conditions, although leaches were quite resistant to chemical agents. In the ponds of urban landscapes we marked the deficit of higher vegetation – substratum for leaches and for various hydrobionts groups, their food base.

Inroduction

Among nature protection problems knowledge and preservation of animal species diversity in conditions of anthropogenic pressure are urgent aims. The present work is concerned with a study of the freshwater leech fauna, peculiarities of their distribution and occurrence in the water reservoirs in urbanized territories of the Middle and Southern Urals.

The Urals are a mountain province located between the East European and West Siberian Plains, they include the Ural mountain system, stretching in meridional direction – from the coast of Kara Sea to dry steppes of Kazakhstan – for over 2,000 km, and 40 to 150 km wide; it consists of the main watershed ridge and lateral ridges divided by wide lowerings. It is partitioned into the Polar, Sub-Polar, Northern, Middle and Southern Urals (Soviet encyclopedic dictionary, 1986). Geological and glacial processes on the Earth have determined the contemporary relief of the Ural Mountains. Relatively low Ural Mountains and Trans-Ural hills occupy its western part, while its eastern part is occupied by the plain, a part of the West Siberian Plain. The Ufa plateau, transacted by river valleys and composed of soluble bed-rocks (limestone, dolomites), is in the south-west. The raised part of the Ural mountains is of the higher axial strip.

In the Ural region, there are thousands lakes, with the total area of 3 million hectares, 35-40 thousand km of large and very large rivers, and storage lakes with an area of more than 600-700 thousand hectares. A plenty of large and small rivers, mountain in their upper reaches and plain in their lower reaches, flows down the western and eastern slopes of the Ural Ridge. Many rivers in the Urals are transected by dams, thus, forming vast factory ponds – specific Ural water reservoirs, an integral part of a typical landscape in the mining and smelting Urals. The majority of ponds have been made for the needs of mining industry in the 18-19 centuries. All of them are depositories for a stock of water in winter and summer. The lakes in the Urals are situated unevenly, they are mostly small and shallow. Most of them are situated in the eastern and in the low places of the West Siberian parts of the province.

The water of most of rivers belongs to the hydrocarbonate class of calcium group characteristic for the fresh water reservoirs of the Ural region (Bessonov, Priverzentsev, 1987). Mineralization of water tends to increase from the north to the south (Water resources…, 2004). The feeding of the water reservoirs is basically atmospheric precipitations and the local surface-water drainage. Four large rivers with vast river basins originate from the territory of the Ural Mountains. These are the Pechora River, from the northern river basin, the tributaries of the Yrtysh River, from the Ob river basin, and the southern rivers – the Ural and the tributaries of the Kama of the Volga-Kama river basin. The Loz'va river basin is typical for the Northern Urals. Among the important water currents of the Irtysh river basin also belongs beginning in the Middle Urals the Iset River and its tributaries from the Southern Urals – the Sinara and the Techa. On the western slope of the Middle Urals the Sylva River belonging to the southern river basin flows (Chikishev, 1966).

In the area of the Ural Mountains, there are the Pechoro-Ilychsky, the Visimsky, and Bashkirsky Nature Reservations.

The Ural region, due to its geographical position and historical reasons, has long ago become a zone of ecological risk caused by a complex of pollutants, (mainly, heavy metals) and radioactive pollution (both natural and artificial) of the environment (Makeyev et al., 2003; Kovalchuk, Yastrebov, 2003).

In the technogenically polluted ecosystems of the Middle Urals there are intensive changes in all the components, circulation of alien to organisms compounds and their inevitable transportation through food chains. This is hazardous both for the normal development and existence of various species, and for the health and reproductive ability of the present and future generations of animals in populations composing various aquatic and terrestrial communities. The most informative parameters during the estimation of ecosystem resistance and stability in technogenic conditions are the changes in biodiversity, species composition and population dynamics (Bolshakov, Kovalchuk et al., 2005).

In the megapolis of Ekaterinburg, the largest industrial center in the Urals, ecological conditions of life are influenced by a developed complex of mining, chemical, metallurgical and machine-engineering industries. Adjacent to the large settlements agricultural grounds experience the influence of industrial pollutants in an area of hundreds square kilometers.

Entering the water basins streams of pollutants are exposed to the physical, chemical and biological influences promoting their auto purification, a part of substances or products of their transformation accumulates in bottom deposits which become centers of secondary pollution. Great quantities of polluting products accumulate in the bottom and can make the water in the rivers and ponds of megapolises the reason for poisoning and diseases. Moreover, the substances which are in the passive form of almost insoluble compounds, can become active in emergency discharge, poison all around and destruct the ecosystem.

As the bottom deposits are concentrators of a significant number of pollutants, bottom organisms, including leeches, are exposed to the maximal influence of toxicants. The role of freshwater leeches in water ecosystems of various types is rather significant. They represent an important link in food chains. Predators and blood-suckers for various groups of hydrobionts, the leeches themselves are the prey for many species of fish, desmans, otters, minks, etc. Some leech species have a "sanitary" role in water reservoirs, as they eat carrion. Besides, theses worms are the additional and intermediate hosts of many parasites.

In the process of evolution, macrozoobenthic communities have developed a series of mechanisms for the preservation of their structure under the influence of the environmental factors. Bioaccumulation, particularly, can be regarded as one of the possible reactions of some representatives of macrozoobenthos to lake pollution.

Leeches, predators and blood-suckers of numerous groups of hydrobionts, are on the top of the food pyramid and owing to their ecological and physiological features are capable of active and significant accumulation of heavy metals transmitted through the food chain (Chernaya, Nokhrina, 2004).

The purpose of the present research was to give a comparative characteristic of the quantitative and qualitative structure of the hirudinean fauna in the water bodies of various type (rivers, lakes, ponds and storage lakes) in the Middle and Southern Urals, and to study the features of accumulation of heavy metals in the tissues of leeches, inhabiting the biotopes in urbanized and natural landscapes.

Materials and Methods

1. Characteristic of Background Territories and Technogenically Polluted Areas

The following 14 water bodies have been chosen for this study, including those situated in the territory of the large industrial center – Ekaterinburg city (the Iset River, the Shartash Lake, the Nizhne-Isetsky and Verkhne-Isetsky ponds), and as in other localities of Sverdlovsk province (the Iset River in the Iset' settlement and in Aramil town; the Pyshma River in Sadovy settlement; the Sysert River in the Sysert town; the Tavatui lake in Tavatui settlement; the ponds in Revda town, in Sysert settlement, in Cherdantsevo settlement; the water reservoir in Verkhny Tagil town). The characteristics of the studied water bodies located on the disturbed by anthropogenic influence territories are given below.

The Iset is the main river in Ekaterinburg, being a left tributary of the Tobol River (the Ob river basin). The total length is 606 km, from the upper reaches to the city – 20 km, within the limits of the city – 35-40 km (together with ponds), and in Sverdlovsk province – 202 km. The average depth of the river is 2.5 m, maximum – 5 m). In its upper part the Iset River accepts a series of tributaries: the Reshetka and the Svetlaya rivers up the city, the Aramilka, Bobrovka, Sysert Rivers down the city. At present, the Iset River has the following tributaries within the territory of Ekaterinburg: the Patrushikha – on the right side, the Olkhovka, the Chernaya and the Istok rivers – on the left. Side According to the data of the State Service for the Environmental Pollution Monitoring, the Iset is the most heavily polluted river in the territory of Russia for some years. The regulated upper streams with the help of ponds and water reservoirs have an imported influence upon the formation of the chemical composition of the water. The Iset is chronically polluted by nitrogen compounds, organic substances, mineral oil, heavy metals (State Report..., 2005). Unformed river grounds are typical for the polluted sites of this river. The part of the river from its sources to the Verkh-Isetsky reservoir (25 km) and the reservoir itself are not polluted with industrial sewage, as they provide the centralized water supply for Ekaterinburg city. The width of the river here is 8-30 m, the current rate – 0.3-0.8 m/s. the depth – 0.5-1.0 m, the bottom is mud sand (Sokolova, 1989).

Within the territory of Ekaterinburg Iset River is "shut" in four places. The first dam on its water course is the one of the Verkh-Isetsky pond. The Verkh-Isetsky pond is the largest artificial reservoir in the city, with the water area of about 14.5 km^2; its 10 km on the Iset. The greatest depth by the dam and along the former fairway of the river is 8-10 m. The hollow of the pond is incorporated in the granites of the Verkh-Isetsky mountain-mass, separate high parts of which form hills approaching the reservoir, and some high isles. The shore line of the pond is alteration of shallow and higher peninsulas. In the warm season the Verkh-Isetsky pond is a favorite recreation zone for citizens of Ekaterinburg, therefore the shore of the reservoir experiences essential anthropogenic loads resulting in the suppression of the aquatic flora and fauna.

The Nizhne-Isetsky pond is an artificial reservoir (one of the four within the boundary of the city), which extends along the Iset River, with flat shores. Its length is 4 km, the water area – about 3 km^2, the average depth – 2.5 m, maximum – 3-3.5 m by the dam. Some time ago, the Nizhne-Isetsky pond was clean enough, local people used its water for economic needs and for fishing. However, after the collector of the city drain began to discharge sewage into the pond, it started to get polluted. The aquatic flora in the reservoir was lost almost completely, the plankton and fish are presented poorly (Arkhipova, 2001).

The Shartash Lake is located in the east suburb of Ekaterinburg, surrounded by low hills, formed by biotite granites, reaching greatest height on the east shore. The lake is bean-shaped, with the bulging side turned to the east. It extends 4 km from the north to the south, 2.0 – 2.5 km – from the west to the east, the water area is 7.0 km^2. The contemporary shore line of the lake, extending about 12 km, is little broken. Its bottom slopes to the center. Most of the bottom is covered with dark-olive colored silts (sapropel), and rubbish. Presently the lake hollow, with its general capacity of 46 million m^3, is almost half covered with silt. The average depth of the lake is 3m, maximum 4.7 m in the center. The shallow water (depth less than 2 m) occupies more than 30%, the depth less than 1 m – 28% of the total water area. Due to vast shoal areas, the water in the lake in summer time gets strongly warm, with the temperature reaching by the shores in the warmest days up to 26-28^0 C. That is why the lake is heavily overgrown, the vegetation dies off rapidly and accumulates at the bottom (Arkhipova, 2001). Earlier, the lake was connected with the Iset River basin by a channel but presently has no runoff. The shores of the lake are dry and woody, a big Shartash village is situated on the northern shore over two centuries. The numerous holiday houses, summer residences and gardens have appeared on the western shore of the lake during the last two decades. As a result, the lake experiences a significant anthropogenic influence (Shlyapnikov et al., 1990).

In summer time, the water in the Shartash Lake contains up to 72 mg/l of chlorides. Salty ammonium, nitrites and nitrates, and substances of organic origin are found too. Besides, the deficiency of the dissolved oxygen (6.4 mg/l – in summer and 4.5mg/l – in winter) is recorded in the Shartash Lake. This water body belongs to the eutrophic type (Piskunov et al., 1984) and is characterized by a high biological productiveness. The Surplus of the organic substance is caused, first of all, by the intra-reservoir processes related to the transformation of biogenous elements. The water quality evidences of a weak pollution of the lake; thus, its ecological condition remains rather stable. This has resulted from the high self-clearing ability of the reservoir due to highly developed zooplankton and zoobenthos –consumers of the primary production (Lugas'kov et al., 2002).

The Kalinovsky Pond is located in the eastern suburb of Ekaterinburg City. Till now this small artificial reservoir fed by underground waters was characterized by high species diversity both of the high aquatic vegetation and of many groups of hydrobionts. However, commercial and amateur fish breeding (introduction of some fish species, clearing the shallow zone of macrophytes) has led to fatal consequences: the ecological equilibrium is disturbed, and this reservoir is subject to irreversible processes of degradation. According to the authors' observations, the total absence of molluscs, larval insects, leeches and other groups of invertebrates is recorded in the pond today.

The Sysert River is the biggest right tributary of the headwaters of the Iset, its length is about 80 km, the average depth – 3 m. Beginning from a small Sysertskoye lake, it has flat boggy shores, a rather narrow channel and silty bottom. The sources of the Sysert and the middle part of its valley are incorporated in the granites of the Aramil-Sysert' mountain – mass, and its lower part – in the sedimentary bed-rocks. In its middle and low parts the Sysert flows in a low hilly plain, forming curves, and accepts some small tributaries – the Gabievka, the Kamenka and the Cheremshanka – on the left, the Raskuikha, the Shaitanka and others – on the right side. From its sources, through Verkhnaya Sysert' to Cherdantsevo, the Sysert flows to the northeast, and then turns to the southeast, and having made a great curve, interflows with the Iset where the settlement of Dvurechensk is situated.

In the Sysert River the oxygen regime is normal, the pollution level is an average one (State Report..., 2005). We have studied the leech fauna in the Sysert River basin within the limits of Sysert' settlement, and in the ponds in the Sysert' and Cherdantsevo settlements.

The Pyshma River in the territory of Sverdlovsk province is 517 km long, and about 80 km – in the vicinity of Ekaterinburg, and in its territory the river forms the city northern border of over 15-20 km (up to the mouth of the Baltymka River). Near Ekaterinburg the Pyshma River accepts some small tributaries: the Kamyshenka, Kalinovka, Baltymka, Beryozovka, Shilovka, Cheremshanka, Stanovlyanka. The first kilometers from its source the

Pyshma flows in a peat bog, its channel is hardly distinguished from the ditches made for peat extraction. After the bogs the river becomes wider but its shores remain flat.

A rather tense ecological situation is observed in the upper reaches of the Pyshma River where communal and industrial sewage from Ekaterinburg, Verkhnaya Pyshma, Berezovsky towns are discharged. Heavy metals, compounds of nitrogen, mineral oil are found in concentrations considerably exceeding maximum concentration limits. Besides chronic pollution of the Pyshma River in its upper-reaches with arsenic (State Report..., 2005) is noted. Significant deficiency of oxygen is also observed. Water pollution index for this reservoir is higher than for other reservoirs listed (Hydrological characteristics..., 1989).

The Verkhne-Tagilskoye storage lake, founded in 1958 on the Tagil River as a reservoir-cooler, is located in the east part of the Middle Ural Mountains, within the boundaries of Verkhny Tagil town. Its water area is 3 km^2, the average depth – 3.8 m, it is 4 km long and 0.8 km wide. The forest-covered area in the range of the Verkhne-Tagilskoye water reservoir is 66%, the bogging area is 8%, no lakes in the watershed. The Tagil River has basically snow filling (60-70% of the annual). Concerning the thermal balance, the Verkhne-Tagilskoye reservoir is in the category of reservoirs with strong overheating, as the water temperature is permanently 6^0C higher than in natural reservoirs. The difference in water temperatures at the inlet of the Power Station and at the outlet to most frequently 9^0, in the range of $6^{\ 0}$ to 12^0C. In winter, the temperature of felling water exceeds 18^0C, and in summer it is often above 30^0C, while in natural water bodies during the warm season it is 10-12^0 lower. The Verkhne-Tagilskoye reservoir does not freeze in winter.

The decrease in algae species composition in the Verkhne-Tagilskoye reservoir is an evidence of pollution. The quantity of protococci, diatoms, desmids species has decreased, euglens and other disappeared. Instead of oligosaprobic *Synedra ulna* var. *danica*, indicator of clear water, *Melosira granulata* var. *granulata* – typical beta-mesosaprobic species – began to develop. The value of saprobic index in the Verkhne-Tagilskoye reservoir varies from 1,9 up to 2,5 i.e. the reservoir is a beta-mesosaprobic type approaching alpha-mesosaprobic type (i.e. stronger pollution). Simultaneously high oxygen deficiency is observed (Vasil'chikova et al., 1989).

As a control we have studied 8 water bodies located in anthropogenically undisturbed territories: the Sulem River and the Sulemskoye reservoir in the Visimsky Nature Reservation (the Middle Urals), the Bol'shoye Miassovo, Small Miassovo, Bol'shoi Tatkul' lakes in the Ilmensky Nature Reservation (the Southern Urals), the upper reaches of the Iset River (up Ekaterinburg city) and two water reservoirs under insignificant anthropogenic pressure – the Beloyarskoye and the Dvurechenskoye (the Middle Urals).

The Ilmensky Reservation in Chelyabinsk province extends to 60 km from the north to the south, of 6-18 km wide, the total area is 50,200 hectares. It is a typical mountain land region. It includes the Ilmen ridge separated from the general system of the Southern Ural Mountains by the Miass River valley, and by the east foothills of the Ural mountains with their complex relief. The reservation itself is located on the border between the forest and the forest- steppe. The most part of the reservation is covered by woods (78% of the total area), lakes occupy 15,1% (Ilmensky Reservation, 1991).

The Bol'shoye Miassovo, a deep foothill lake (maximum depth – 23 m, average depth – 11 m), has a broken shore line and a complex form of the hollow. Its grounds and shore forms are variable – from stony, stone-sandy to peat and oozy. Macrophytic vegetation is well developed in the shallow gulfs and is rather poor in the narrow strip of the littoral zone on the open shores. There are quagmires in the gulfs. The lake belongs to the mesotrophic type with signs of eutrophy in the gulfs. It is characterized by a high species diversity of the high aquatic vegetation and of many groups of hydrobionts.

The Bol'shoi Tatkul' is a shallow foothill lake (maximum depth – 6.5 m, average depth – 4,5 m), with a leveled shore line and smoothed section of the bottom. Its shores are boggy, with quagmires, the grounds are mainly silty. The Bol'shoi Tatkul Lake is almost completely surrounded by ancient forested quagmire. Almost in the whole bottom area macrophytes occur; the lake of the eutrophic type.

All the lakes mentioned above are located in the territory of the reservation. Fall-out of pollutants with atmospheric precipitations is an anthropogenic factor.

The Maloye Miassovo Lake forms a common hydrological system with the Bol'shoye Miassovo Lake. Both are connected with a channel. One of it parts is located in the hilly foothill zone, the other part is in the flat Trans-Urals, in a forest-steppe zone. It has moderate depths (maximum – 8 m, average – 4.5 m). Its foothill part is similar to the Bol'shoye Miassovo Lake, the flat part has a flat shore, small depths, an even bottom. It is subjected to the influence of anthropogenic factors related to the presence in the shore zone of numerous holiday-houses and settlements. Natural fluctuations in the water level in the Bol'shoye Miassovo and the Maloye Miassovo are accompanied by fluctuations due to regulatory effects of the dam. The Bol'shoye Miassovo Lake is distinguished by the highest phytocoenotic richness. It is poorer in the Maloye Miassovo lake, and the least quantity of plant formations is found in the Bol'shoi Tatkul Lake. In the latter, macrophytes *Ceratophyllum demersum* L. and *Stratiotes aloides* L. dominate – an evidence of an increased eutrophy level (Weisberg, 2007).

The Visimsky Biospheric Reservation is located in the southern taiga subzone of the Middle Urals, on the easternmost boundary of Europe, on the watershed of the eastern and western macroslopes of the Ural-Tau ridge (a mountain part of Middle Urals mountains), where it almost completely covers the upper streams of the Sulem River (the right tributary of the Chusovaya River) and, partly, of the Tagil River. The Visimsky Reservation is one of the best studied nature reservations in Russia concerning species diversity of fungi, plants and animals. The fauna of invertebrates in this reservation numbers 1743 species and needs a farther thorough studying (Marin, 2006). The fauna of ground invertebrates in the Sulem River basin in the territory of the Visimsky Reservation numbers about 168 species and forms. The basic part in the formation of the numbers and biomass of the ground animal communities is played by larval chironomids, caddis-flies, may-flies, stone-flies and molluscs. High quantitative values of the development of zoobenthos are characteristic for the moss periphytic cover of the stone-pebbly grounds in the of Sulem River headwater (Stepanov, 2001).

The Beloyarskoye water reservoir is located 60 km to the east of Ekaterinburg. This reservoir was formed in 1959-1963, due to the Pyshma River regulation, at a distance of 75 km from its source. The extent of this water body is about 20 km, the width – about 3 km. The depth along the Pyshma River fairway is 15-20 m, the average depth –about 8-9 m. The water area of the reservoir is 47 km^2. Small rivers the Pushkariha, Cheremshanka, Chernaya, Marja, Lipovka, etc. run into this reservoir. The shores are mainly flat, stony here and there, generally occupied by forests, with mosaic occurrences of shrubby bogs and pasturable meadows. Concerning quality of water the reservoir is attributed to the hydrocarbonate calcic type, with an average level of mineralization and alkalescent reaction of the environment.

The Beloyarskoye water reservoir is now in the initial stage of its formation, so aquatic plants can be more often met in the gulfs than in its central part. Among 29 registered species of macrophytes, pondweeds, cladophores, Canada water weed, hornwort submerged and some other species prevail. The ichthyofauna in the reservoir is presented basically by the roach, bream, northern pike, mirror carp; the tench, crucian carp, river perch and sander are less numerous. In the phytoplankton green algae (chlorococci, desmidians, volvoxes) have a greater number of species; diatoms, blue-green, euglenophytic, pyrophitic and other groups of algae are less spread. Crustaceans (cladocerans and copepods) are the bulk of the zooplankton, rotifers are met in smaller numbers. The prevailing type of ground deposits in the reservoir is silty sapropel; the flooded, sandy-oozy, oozy-sandy and sandy grounds occur, too. 7 km to the north from the dam of the reservoir-cooler the Beloyarsk Nuclear Power Station is located on the left shore. Slightly radioactive liquid's discharged to the reservoir-cooler. Lower is the water-waste (warm) channel, by which the water enters the reservoir after passage through the cooling system. In the heating zone the water temperature in the summer period is averagely 6-7^{0}C higher than outside it. Presence of a zone with a higher temperature of water and the alkalescent reaction of the environment create favorable conditions for fishes and other hydrobionts. The normal oxygen regime – 10.4 mg/l – is observed in the reservoir (Chebotina et al., 2002).

The Dvurechenskoye storage lake is located at the confluence of the Iset'- and Sysert Rivers. As a rather small water body it exhibits insignificant anthropogenic loads and has a relatively high number of macrophytes and hydrobiont species.

2. Methods

We collected the biological material during the field seasons (May-October, 1999-2006) in the shore areas of 22 water bodies described above. Totally, 260 samples were collected, more than 11 thousand adult individuals of various leeches were caught.

Considering the biological features of this group of organisms (leeches are suckers and spend a significant period of life in an attached position), we collected them by hand from stones, snags and other submerged subjects, usually from the reverse, poorly lighted side. Besides, the leeches were collected from the aquatic plants, mainly, from the axils of leaves and from old, brown, submerged stems.

The quantitative sampling was carried out in the water bodies crowded by the main species in the first decade of August by the standard technique (Mann, 1955). The abundance of each species collected during one hour per a square meter of the bottom was estimated.

The analysis of all samples revealed the dominant species and determined the quantitative and qualitative composition of the leech fauna (percentage) in a particular reservoir. Concentrations of heavy metals (HM) in leech tissues were determined using the method of atomic-absorbtion spectrometry on the spectrophotometer AAS-3 in a flame of propane-butane (Khavezov, Tsalev, 1983) after processing the samples by the "wet" incineration method in a mixture of nitric and chloric acids (Nikanorov et al., 1985). The HM content was expressed in mkg/g of the wet mass.

To estimate HM accumulation in leech tissues, we chose the most widespread in the Ural ecosystems toxicants – copper, lead and cadmium (Kovalchuk et al., 2002). Essential microelements – copper (Cu) and zinc (Zn) – can cause toxic reactions under certain conditions and at higher (Cu) or lower (Zn) concentrations. Exogenous microelements – lead (Pb) and cadmium (Cd) – have high toxicological characteristics and have injurious, pathological effects on organisms.

The experimental part of the research was made on adult individuals of three species of leeches. They were delivered to the laboratory the day of their capture. 12 hours after their intestines were dragged out, they were dried with paper filters, then a quantity of one g of their skin-muscular tissue was taken and dried at 50-60^{0}C in the thermostat to dry condition. To estimate HM content in the skin-muscular tissue of the leeches, 75 samples were prepared and analyzed. Depending on the individual body masses, the following numbers were used in the experiments: 25 individuals of predatory leech *Haemopis sanguisuga* (L., 1758), family Hirudinidae; 210 individuals of blood-sucking leech *Glossiphonia complanata* (L., 1758), family Glossiphoniidae; 317 individuals of predatory leech *Erpobdella octoculata* (L., 1758), family Erpobdellidae.

Statistical processing of the results was made using programs STATISTICA 6.0, with subsequent estimation of distinctions at $p \leq 0.05$.

Results

The results of the faunistic investigations have shown that 11 species of freshwater leeches belonging to 4 families occur in the water bodies of the Middle and Southern Urals

I. Family Glossiphoniidae Vaillant, 1890

1. *Protoclepsis tessulata* (O. F. Müller, 1774) – blood-sucker parasite of the waterfowl;
2. *Protoclepsis maculosa* (Rathke, 1862) – blood-sucker of the waterfowl;
3. *Hemiclepsis marginata* (O. F. Müller, 1774) – blood-sucker of fishes and amphibians;
4. *Glossiphonia complanata* (L., 1758); blood-sucker of molluscs and larval insects;
5. *Glossiphonia concolor* (Apathy, 1888) – blood-sucker of molluscs and larval insects;
6. *Helobdella stagnalis* (L., 1758) – predator on small invertebrates;

II. Family Ichtyobdellidae Leuckart, 1863 – The Fish Leeches

7. *Piscicola geometra* (L., 1758) – fish parasite, a typical oxyphilic species.

III. Family Hirudinidae Whitman, 1886

8. *Haemopis sanguisuga* (L., 1758) – glutinous predator on water invertebrates and small vertebrates (tadpoles, fry fish); amphibiotic life-style.

IV. Family Erpobdellidae Blanchard, 1892

9. Erpobdella octoculata (L., 1758);
10. Erpobdella testacea (Savigny, 1822);
11. Erpobdella nigricollis (brandes, 1900). Three species of leeches are predators on small invertebrates.

The comparative analysis has shown that fauna of leeches in urbanized territories is poorer than in natural landscapes, both quantitatively and qualitatively (tables 1, 2).

Table 1. Species composition of leeches in water reservoirs in the urbanized territories of the Urals

Reservoirs	***Glossiphonia complanata***	***Glossiphonia concolor***	***Helobdella stagnalis***	***Erpobdella octoculata***	***Erpobdella nigricollis***	***Erpobdella testacea***	***Haemopis sanguisuga***
Iset River (Ekaterinburg)	+	-	+	+	+	-	+
Iset River (Istok settlement)	+	+	+	+	-	-	-
Iset River (Aramil town)	-	-	+	+	-	-	-
Sysert River (Sysert settl.)	-	-	+	+	-	-	+
Pyshma River (Sadovy settl.)	-	-	-	+	-	-	+
Tavatui lake (Tavatui settl.)	+	-	+	+	+	-	-
Shartash Lake (Ekaterinburg)	+	-	+	+	+	+	-
Verkh-Isetsky Pond (Ekaterinburg)	+	-	+	+	-	-	-
Kalinovsky Pond (Ekaterinburg)	+	-	+	+	-	-	+
Nizhne-Isetsky Pond (Ekaterinburg)	+	-	+	+	-	-	-
Revdinsky Pond (Revda)	-	-	-	+	-	-	-
Sysert Pond (Sysert settl.)	+	-	-	+	-	-	-
Cherdantsevo Pond (Cherdantsevo settl.)	+	-	-	+	-	-	-
Verkhne-Tagilskoye reservoir (Verkhny Tagil town)	+	-	-	+	+	-	-

Table 2. Species composition of leeches in reservations and in anthropogenically undisturbed territories of the Urals

Reservoirs	***Glossiphonia complanata***	***Glossiphonia concolor***	***Helobdella stagnalis***	***Protoclepsis tessulata***	***Protoclepsis maculosa***	***Erpobdella octoculata***	***Erpobdella nigricollis***	***Erpobdella testacea***	***Piscicola geometra***	***Hemiclepsis marginata***	***Haemopis sanguisuga***
Iset River (upper sources)	+	+	+		-	+	+	+	+		+
Belo-yarsk reservoir (Bio-physical station of IPAE)	+	+	+	+	-	+	+	+	+	+	+
Dvure-chensk reservoir (biosta-tion of USU)	+	+	+		-	+	+	-	-	-	+
Sulems-koye reservoir (Visim Reser.)	+	+	+		-	+	+	+	-	-	+
Sulem River (Visim Reserv.)	+	-	+		-	+	+	+	-	+	+
Bol'-shoye Miassovo Lake (Ilmen-sky Reserv.)	+	+	+	+	+	+	+	+	+	+	+
Maloye Miasso-vo Lake (Ilmens-ky Reserv.)	+	+	+	+	-	+	+	+	-	+	+
Bol'shoi Tatkul Lake (Ilmen-sky Reserv.)	+	+	+	-	+	+	+	+	-	+	+

Thus, in reservoirs of the urban agglomerations seven species of leeches against the background of their low abundance were revealed. The most widespread species in urbanized territories were predatory *E. octoculata* (in 14 water bodies) and *H. stagnalis* (in 11 water bodies). The most numerous species in all reservoirs was *E. octoculata* (39.7% of the total number of collected individuals of seven species), the percentage of *H. stagnalis* was 11.2%. The blood-sucker of molluscs, *G. complanata,* was usual enough for the fauna of leeches in Ekaterinburg and other settlements (it was met in 10 water bodies and made up 6.7% of the total number). *E. nigricollis* was recorded only in four reservoirs, it was more abundant in natural lakes (the Tavatui and Shartash), where it exceed in numbers all other species. During the whole period, *H. sangusuga* was found only in five city reservoirs, not in all samples, it was a scanty species. The predatory leech *E. testacea* and the blood-sucker of molluscs *G. concolor*, its competitor for food, the ecologically plastic *G. complanata,* were found only in one water body in a low number.

On the whole, the low abundance of all leech species has been recorded in all reservoirs located in the territory of towns and settlements: no more than 10 individuals per a square meter per hour. An exception was the Shartash Lake (79 ind./m^2/h and the Verkhne-Tagilskoye reservoir (73 ind./m^2/h) where five leech species were met.

In reservoirs of urbanized landscapes we recorded the deficiency of high aquatic vegetation – substratum for leeches and groups of hydrobionts, and a food resource for the last.

On the contrary, in natural reservations and in anthropogenically undisturbed s territories rich species diversity of macrophytes, various groups of hydrobionts (crustaceans, larval insects, molluscs, fish, amphibians), and the waterfowl, was observed i.e. conditions were favourable for 11 leech species (table 2). As this table shows, the highest species diversity was in lakes in the Ilmensky Reservation – the Bol'shoye Miassovo (11 species), the Maloye Miassovo (10 species), the Bol'shoi Tatkul' (10 species) – and the Beloyarskoye reservoir (10 species). Dominant and mass species in natural water bodies were predators and blood-suckers. Thus in the Bol'shoye Miassovo and the Maloye Miassovo predator *E. nigricollis* was the most numerous species, while in the Bol'shoi Tatkul lake parasite of molluscs *G. concolor* dominated. Against the background of high species diversity of aquatic invertebrates a higher abundance of molluscs was recorded in the Bol'shoi Tatkul lake, compared with the determined Bol'shoye and Maloye Miassovo lakes. This factor, in our opinion, the difference in dominant species of leeches in those Lakes. The Predatory leech *E. nigricollis* prevailed in numbers in the Sulemskoye reservoir (the Visimsky Reservation), too, the blood-sucking *G. concolor* – in the Dvurechenskoye reservoir. The predator *E. octoculata* dominated in other water bodies.

The percentage of the collected individuals of these main species was about the same: *E. nigricollis* – 23.5%, *G. concolor* – 21.7%, *E. octoculata* – 21.3%. The high enough abundance of the main leech species (119-201 ind./m^2/h) was found in all control reservoirs.

Predatory leeches *H. sanguisuga, H. stagnalis, E. testacea* were widespread in the control water objects, however, they were considerably lower in numbers: to mass species of leeches. The avian Blood-sucker P. *tessulata* and *P. maculosa* were rather rare and no numerous in the Urals. The fish parasite *P. geometra* was found only in three water reservoirs. This species is typically oxyphilic and very susceptible to chemical pollution of the water environment.

The spectrophotometric analysis has shown, that in the Verkhne-Tagilskoye storage lake located in the Verkhny Tagil town, in the zone affected by copper-smelting works, predatory leeches *H. sangusuga* accumulated maximal concentrations of copper (17.5 $\pm$ 0.53 mkg/g), zinc (269.4 $\pm$ 4.92 mkg/g) and cadmium (0.94 $\pm$ 0.07 mkg/g). which greatly exceeding the background values (figures 1, 2). The raised thermal regime in this reservoir-cooler promoted the more intensive accumulation of toxicants in the tissues of leeches. The highest concentration of lead were revealed in the leeches from the reservoirs of Ekaterinburg: in *G. complanata* in Shartash Lake – 3.26 $\pm$ 0.3 mkg/g. in *H. sanguisuga* in Iset River – 2.63$\pm$ 0.26 mkg/g, evidently due to high transport and technogenic loads in the large industrial centre (figure 2).

It is interesting, that concentrations of heavy metals in tissues of leeches from nature reservations did not differ significantly from the values in urbanized territories.

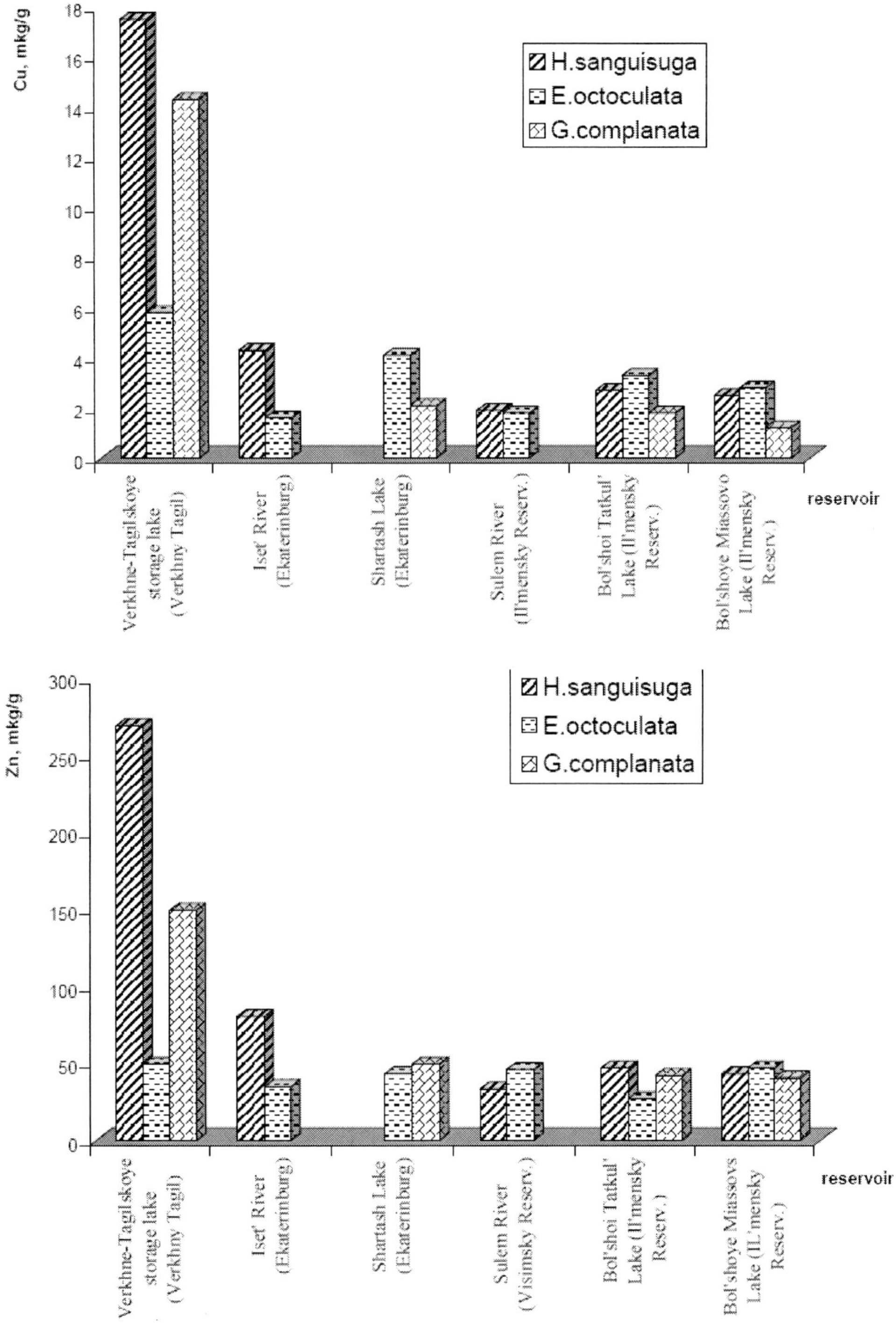

Figure 1. Concentrations of essential microelements – copper (Cu) and zinc (Zn) – in the tissues of leeches inhabiting water reservoirs in the Middle and Southern Urals.

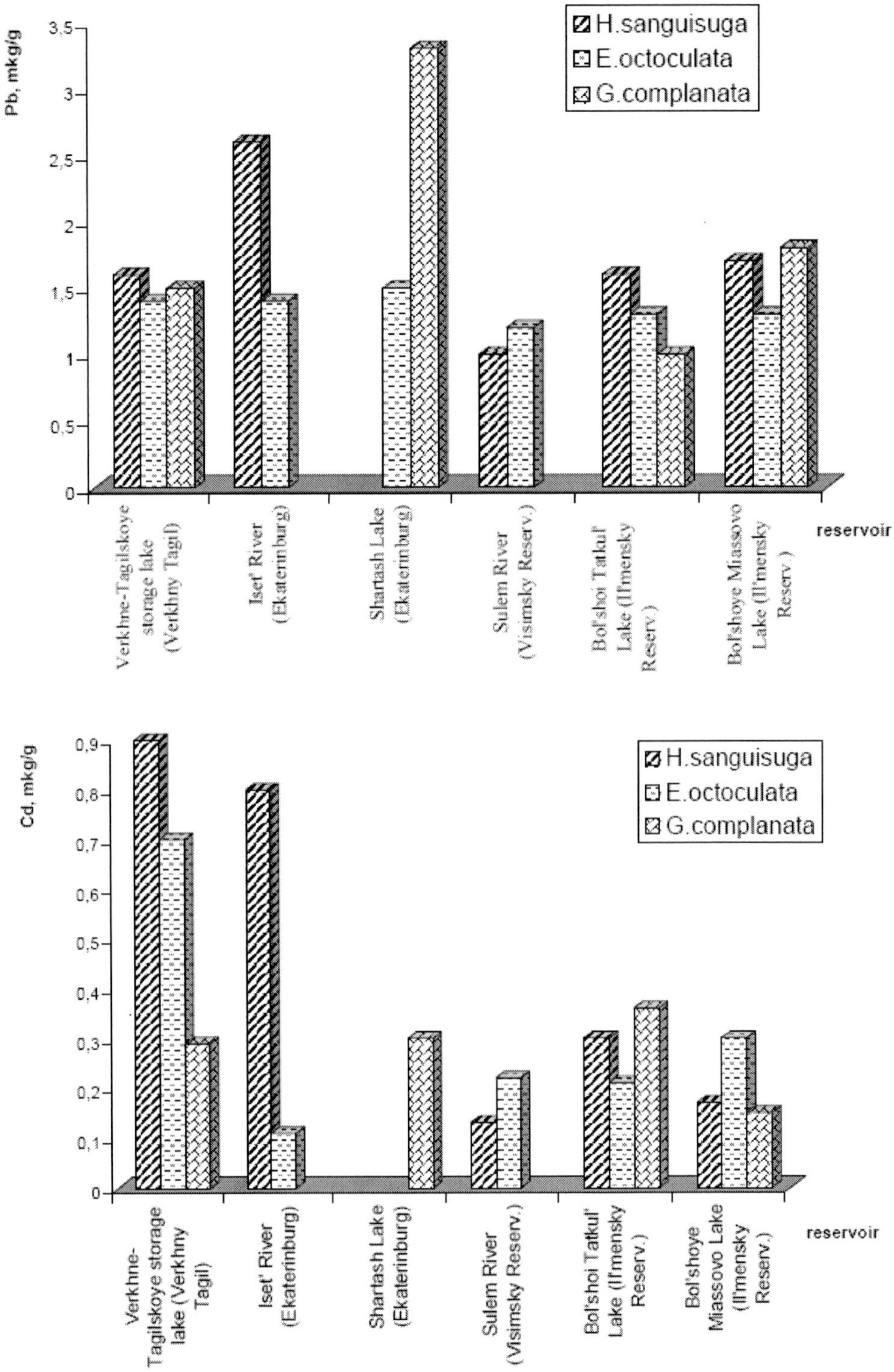

Figure 2. Concentrations of exogenous microelements – lead (Pb) and cadmium (Cd) – in the tissues of leeches inhabiting water reservoirs in the Middle and Southern Urals.

For example, the contents of copper in *G. complanata* from the Shartash Lake in the territory of Ekaterinburg and from the Bol'shoi Tatkul lake in the Ilmensky Reservation (t_{1-2} = 0.21 < t_{st} = 2.31, $p < 0.05$) did not differ significantly. No differences in accumulation of zinc were not found in *E. octoculata* from the Iset River in Ekaterinburg and from Sulem River reserve (t_{1-2} = 1.24 < t_{st} = 2.31, $p < 0.05$). The fact that evidences the level of technogenic pollution was not a determinative factor of oppressive condition of leech fauna in urbanized territories.

Laboratory toxicological experiment (Chernaya, Kovalchuk, 2002), showed, that representatives of the main leech species inhabiting water ecosystems of the Urals are resistant to the action of toxicants (Cu, Zn, Cd) (table 3). In the experiment, the Ural leeches remained alive at concentrations of Cu and Cd of 0.01 mg/l, Zn of 0.03 mg/l, while the leeches from the background territories died, and the change of the static state to the dynamic one (CSSD) was recorded in all species in the last case (table 3).

Table 3. Toxicological significance of CSSD test-reaction in the leeches of natural populations from the impact (Sverdlovsk Province) and background (Tambov Province) territories

Species	HM	HM-concentr. (mg/l)	CSSD, % (Sverdlovsk Prov.)	CSSD, % (Tambov Prov.)	Death time (days) (Sverd-lovsk Prov.)	Death time (days) (Tambov Prov.)
E. octoculata	Cu	0.01	0	60	-	9
	Zn	0.03	20	40	-	-
	Cd	0.01	40	60	-	6
E. nigricollis	Cu	0.01	20	40	-	9
	Zn	0.03	0	20	-	-
	Cd	0.01	40	60	-	16
G. complanata	Cu	0.01	0	20	-	10
	Zn	0.03	0	40	-	13
	Cd	0.01	0	40	-	8
G. concolor	Cu	0.01	0	20	-	12
	Zn	0.03	20	60	-	18
	Cd	0.01	20	40	-	9

This resistance can explain the presence of *E. octoculata, H. sanguisuga, H. stagnalis* in the water bodies affected by high technogenic loads. However, their low abundance suggests, that water pollution affected in directly, through changing life conditions and food resources (deficiency of macrophytes and zoobenthic organisms). Although *G. complanata* and *G. concolor* were a highly resistant to the effect of heavy metals in the experiment, in the water bodies subjected to anthropogenic pressure, these species were absent or single individuals were met gastropods – their main food supply for glossiphoniids.

Conclusion

Based on the data of the research conclude, that urbanization has fatal effects on the specific and number structure of the leech fauna, that mostly predatory species survive in water reservoirs of towns and settlements. This is evidently, due to extinction of many species of hydrobionts (molluscs, amphibians, fishes – food objects for blood-sucking leeches), of higher aquatic vegetation in conditions of anthropogenic pressure in water ecosystems in urbanized territories.

However, there are many fundamental problems to be solved in this field of investigation. The obtained results are to be complement to and the data received by other researchers allow to estimate biodiversity and ecology of hydrobionts in natural and anthropogenic landscapes.

References

Arkhipova N. P. *The natural sights of Ekaterinburg and its vicinities*. Ekaterinburg: Akva-Press. 2001. 226 p. (In Russian).

Bessonov N. M., Priverzentsev Yu. A. Piscicultural hydrochemistry. Moscow: *Agropromizdat*. 1987. 159 p. (In Russian).

Bolshakov V.N., Kovalchuk L.A., Satonkina O.A., Yastrebov A.P. Energy metabolism and accumulation of heavy metals in tissues of mouse-like rodents in natural and technogenic landscape of the Ural region, *Herald of the Urals Medical Academical Science*. 2005. № 4. P. 12-18. (In Russian).

Chebotina M.Ya. , Guseva B. P., Trapeznikov A.V. Plankton and its role in radionuclids migration in the reservoir-cooler of the atomic station. Ekaterinburg: *Ural Div. RAS*, 2002. 171. p. (In Russian).

Chernaya L.V., Kovalchuk L. A. Adaptive features of natural populations of some species of a class Hirudinea in the conditions of anthropogenic transformation, Biota of mountain territories: history and contemporary condition. *Proc. of young sci. conf. Ekaterinburg*. 2002. P. 270-280. (In Russian).

Chernaya L. V., Nokhrina E. S. Specificity of accumulation of heavy metals in three species of leeches inhabiting the artificial water bodies of Sverdlovsk Province Ecology of industrial region and ecological education. *Proc. sci. conf. Nizhniy Tagil*, 2004. P.180-183. (In Russian).

Chikishev A.G. Physical and geographical division into districts of the Urals. *Problems of physical geography of the Urals*. 1966. P. 7-84. (In Russian).

Ilmensky Reservation. Chelyabinsk: *South Ural Publ.* House. 1991. 156 p. (In Russian).

Khavezov I., Tsalev D. Atomic-absorbic analysis, Trans. from Bulgarian by G.A, Sheinina, S. Z. Yakovleva. Leningrad: *Chemistry*. 1983. 144 p. (In Russian).

Kovalchuk L. A., Satonkina O. A., Tarkhanova A. E. Heavy metals in the environment of the Middle Urals and their influence upon an organism, *Ecology*. 2002. №5. P. 358-362. (In Russian).

Kovalchuk L.A., Yastrebov A. P. Ecological physiology of small mammals of the Urals. Ekaterinburg: *NISO Ur. Dept. RAS*. 2003. 202 p. (In Russian).

Lugas'kov A. V., Bogdanova E. N., Stepanov L. N., Yarushina M. I., Lugas'kova N. V. The results of ecological monitoring of Shartash Lake. Ecological safety of the Urals. Proc. sci. conf., *Ekaterinburg*, 2002. P. 152. (In Russian).

Makeyev O. G., Yastrebov A. P., Izmailov I. H., Tarasevich A. A., Kostyukova S. V., Buraev M. E., Vorob'ev A. V. Contribution of natural radioactivity to a health of inhabitants of MO Karpinsk // *Bulletin of USMA*. Ekaterinburg, 2003. № 11. P. 50-58. (In Russian).

Mann K. H. The ecology of the British freshwater lake leeches. *Journ. Animal Ecol.*, 1955. Vol. 24. № 1. P. 98-119.

Marin Yu. F. Investigations and ecological monitoring in the Visimsky Reservation in 2001-2005 // *Ecological investigations in the Visimsky Reservation*. Ekaterinburg, 2006. P.14-20. (In Russian).

Nikanorov A. M., Zhulidov A. B., Pokarzhevskiy A. D. Biomonitoring of heavy metals in freshwater ecosystems. Leningrad: *Hydrometeoizdat*. 1985. 144 p. (In Russian).

Piskunov L. I., Oshchepkova A. N., Treyger S. I., Lyubimova S. A., Karagodina I. V. Some ecological features of water reservoirs of the Middle Urals, *Hydrobiological Journ*. 1984. Vol. 20. № 5. P. 54-61. (In Russian).

Shlyapnikov D. C., Demchuk I. G., Okunev P. V. Mineral components of the ground deposits in the lakes of the Urals. Sverdlovsk: *Ural Univ. Publ.* House. 1990. P. 104. (In Russian).

Sokolova G. A. Biology of chironomids of the upper reaches of the Iset River, *Hydrobiological characteristic of water reservoirs of the Urals: Proc. sci. works*. Sverdlovsk, 1989. P. 113-116. (In Russian).

Soviet Encyclopedic Dictionary. *Moscow: Soviet Encyclopedia*. 1986. 4-th ed. 1600 p. (In Russian).

State Report on a condition of environment and influence of environmental factors upon the health of population of Sverdlovsk province in 2003. Ekaterinburg, 2005. 255 p. (In Russian).

Stepanov L. N. To the fauna of ground invertebrates of Sulem River and its tributaries, *Research of standard natural complexes of the Urals*. Ekaterinburg, 2001. P. 200-201. (In Russian).

Vasil'chikova A. P., Popov A. N., Berdysheva G.B. Phytoplankton as indicator of water quality in the water reservoirs-coolers in the Urals // *Hydrobiological characteristics of water reservoirs of the Urals: Proceedings of scientific works*. Sverdlovsk: Ur. Dept. AS USSR. 1989. P. 13-22. (In Russian).

Water resources of Sverdlovsk province. Under sci. ed. N. B. Prokhorova; FGUPRosNIIVH. Ekaterinburg: Publ. House AMB, 2004, 432 p. (In Russian).

Weisberg E. I. Specific and phytocoenotic diversity of macrophytic vegetation of polytypic lakes of the Chelyabinsk province. *Proceedings of the Chelyabinsk Science Center*. 2007. Vol. 2 (35). P. 141-146. (In Russian).

In: Focus on Urbanization Trends
Editor: Laura M. De Smet

ISBN: 978-1-60456-109-8

Chapter 6

Sustainable Rates of Groundwater Extraction in an Urbanized Plain, Central Japan

Adrian H. Gallardo[1*], Atsunao Marui[1], Shinji Takeda[2] and Fumio Okuda[2]

[1]AIST, Geological Survey of Japan、 Higashi 1-1-1, Central 7, Tsukuba 305-8567, Japan

[2]Hytec Co. Yodogawa-ku, Miyahara 2-11-9, Osaka 532-0003, Japan

Abstract

Land subsidence due to excessive extraction of groundwater is a major environmental problem that has affected for several decades both the urban centers and the surrounding peri-urban farmlands scattered within the Nobi Plain, Central Japan. Even though strict regulations successfully reduced the subsidence rates, the threat is still far from being eliminated. The plain is one of the most important economic centers of Japan, sustaining an intensive agricultural and industrial production. Furthermore, its population density is among the highest in the country, averaging more than 1,000 persons per km^2. The recent economical growth of the region is expected to be accompanied by new developments and a higher demand for water supply therefore, authorities are compelled to update management strategies to satisfy the increasing demand without causing undesirable effects on the environment. Knowledge of the groundwater availability becomes an essential requisite for the establishment of these policies. A numerical simulation was applied to understand the groundwater flow in the region. The resulting model was then used to determine the safe yield of the aquifers under different management options. The analysis concentrated on the Aburashima site, as this area faces the possibility of development in the forthcoming years. Calculations indicated that withdrawal from the shallow aquifer is limited. In addition, a decline in piezometric levels reduced the groundwater availability in summer. Spreading a number of wells around the site constitutes an alternative to increase the pumping volume. Higher quantities and lower extraction costs makes the lower aquifer a more reliable source of

* E-mail address: ad.gallardo@aist.go.jp. (Corresponding author)

groundwater. The present study constitutes one of the first steps towards the implementation of an integrated management plan of the water resources in the Nobi Plain, and constitutes a valid reference to update the current legislation regarding groundwater use in the region.

1. Introduction

With an area of about 1800 km^2 and approximately 4.5 million people, the Nobi Plain is the second largest populated region in Japan, only behind the Tokyo metropolis and surroundings. As in many other urban areas in Asia (e.g., Tokyo, Niigata, Osaka, Shangai, Bangkok, Hanoi, Calcutta, Taipei), excessive groundwater abstraction from a series of confined aquifers resulted in compaction of clay interbeds and the consequent land subsidence of the region. The issue dates back to the 1950`s when the area underwent an explosive development following the post-war recovery. The subsidence increased exponentially and registered a peak in 1973 when the elevation at some areas decreased more than 20 cm (Yamamoto, 1984). Land subsidence can have several negative implications, as damage to buildings and roads, groundwater quality deterioration and salt encroachment, decline in storage capacity, change in flow patterns, and localized flooding (Larson et al., 2001). Given the magnitude of the problem, a number of regulations on groundwater pumping were enforced from 1974, effectively bringing the situation under control. As a result, areas subsiding more than 1 cm/yr reduced from 293 km^2 in 1975 to 9 km^2 in 2004 (METI, 2006).

Despite the success of the implemented strategies, the recent economic growth of the region is promoting new industrial and agricultural developments, which are accompanied by an increase in the demand of groundwater. Thus, the management authorities are compelled to meet the consumer requirements while preventing undesirable effects on such a sensitive environment. Aware of these needs, the prefecture of Gifu supported the present work towards the establishment of a modern plan for management of the water resources in the Nobi Plain. Results constitute also a scientific tool useful to update the current legislation on groundwater exploitation in the region. The analysis concentrated on the Aburashima site, southwest of the Plain, as it locates at the tripartite boundary of the Aichi, Gifu and Mie prefectures, all of them involved in the implementation of control policies within the basin. In addition, the site faces the possibility of new projects in the forthcoming years which make it especially susceptible to unacceptable consequences. Understanding the groundwater movement will ensure its proper utilization (Don et al., 2005) therefore, a numerical simulation was applied to obtain a detailed picture of the flow patterns over the region. After calibration, the model was utilized to determine the maximum rates of groundwater feasible to be extracted on a sustainable basis at Aburashima without causing a negative impact into the environment. Finally, some alternatives were evaluated to optimize the water supply from the aquifers.

2. Overview of the Area

The Nobi Plain occupies an area of about 45 km by side located within the prefectures of Gifu, Aichi and Mie, in the central part of Japan. It is surrounded by the Paleozoic basement of the Owari Hills to the north and east, the Yoro Mountains to the west, and the Ise Bay on the south (Figure 1). Development of Nagoya and neighbor cities transformed the area into a major industrial center with several millions of residents. Four large rivers (Ibi, Nagara, Iso and Shounai) and a number of minor tributaries characterize the surface waters of the region. Despite the existence of an adequate channeling and tap system, there is still a heavy dependence on groundwater for public supply. The plain lies in a tectonic basin tilted westward, and filled with variable sediments mainly deposited during the Pliocene and Pleistocene (Figure 2). The main aquifers comprise three layers of river-bed gravel deposited during glacial advances (Uchida et al., 2003). From surface to depth, these aquifers are called the Fist, Second, and Third Gravel Bed, also known as the G-1, G-2, and G-3 respectively. Groundwater is mainly withdrawn from the shallow unit, while water for industrial use is mostly pumped up from the Second Gravel Bed (Yoshida et al., 1991). Since most production wells concentrate in the first and second aquifer, the present study restricted to a depth of 200 m, in coincidence with the base of the G-2. As already stated, the large amount of pumping has caused huge problems of subsidence and salinization, and as result, a large part of the Nobi Plain is below sea level (Yamanaka and Kumagai, 2006).

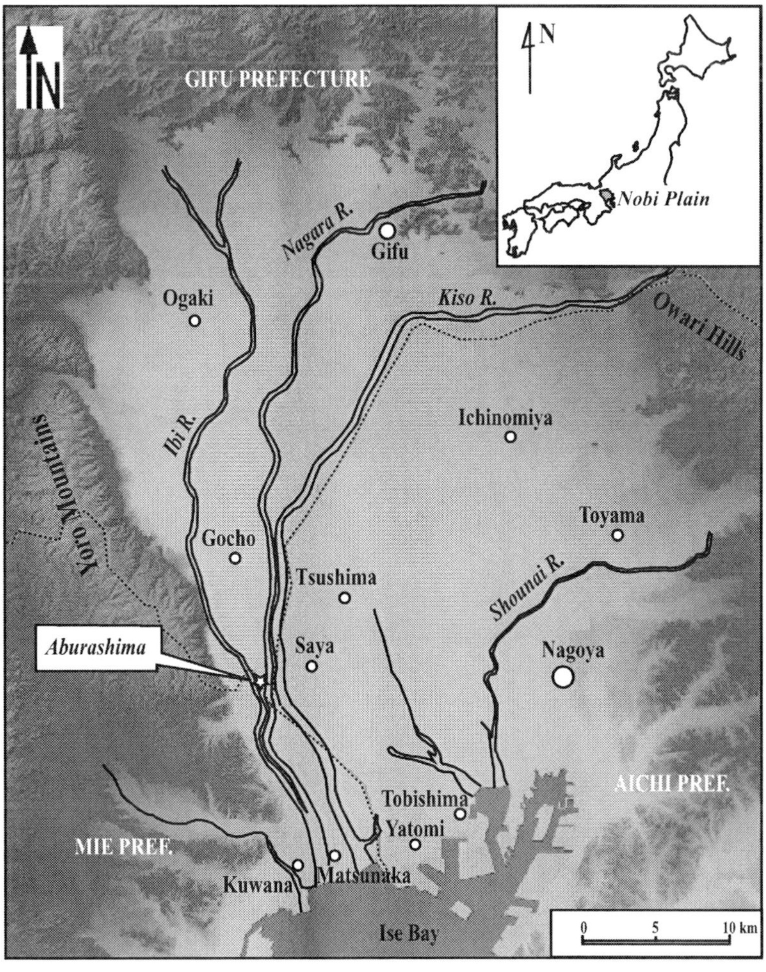

Figure 1. Location map of the Nobi Plain.

Period		Formation	Remarks
Quaternary	Holocene	Nanyo	Alluvium. Transgression
	Pleistocene	Nobi	Last Glacial
		First Gravel Bed	
		Atsuta	Last Interglacial
		Second Gravel Bed	Transgression and Regression
		Ama	
		Third Gravel Bed	
		Yatomi	
Neogene	Pliocene	Tokai Group	Tokai Lake

Figure 2. Summary of stratigraphy and main geologic events in the Nobi Plain district (after Sakamoto et al., 1986).

The aquifers in consideration are separated by a series of silt and clays which limit the vertical movement of groundwater and promote the confining conditions of the aquifers. At the top of the sequence, the Nanyo and the Nobi formations are formed by a thin layer of alluvial sands followed by nearly 40 m of clays with a hydraulic conductivity in the order of 10^{-8} cm/sec. On the other hand, the First and the Second aquifers are separated by the Atsuta formation, deposited in the inner bay and its surroundings during the Last Interglacial and early half of the Last Glacial Times (Sakamoto et al., 1986). Its upper member is composed by 33 m of fine to medium sands intercalated with thin horizons of clay rich in organic matter. In contrast, the lower end is formed almost exclusively by densely packed clays reaching a thickness of about 40 m at the investigation site. Hydraulic conductivity ranges from 2×10^{-4} cm/sec at the top of the formation, to 4×10^{-8} within the deeper clays (Table 1).

The Japan Meteorological Agency maintains updated records of precipitation throughout the country. Data for the period 1986-2005 was used to calculate the mean rainfall within the plain. The highest rates corresponded to the Ogaki station in the north, with an average of 1876 mm/yr. Quantities reduced to 1676 mm/yr at Aichi station in the center of the plain, and reached a minimum of 1596 mm/yr at Kuwana city, near the Ise Bay. There is no information about evapotranspiration so, rates were calculated by the Thornthwaite method (1948). The procedure is based on the assumption that potential evapotranspiration is dependent only upon meteorological conditions, ignoring the effects of vegetation. While this assumption is not exact, the method is reasonably accurate in determining annual values, especially in humid areas (Fetter, 2000). The potential evapotranspiration (ET) is then calculated by the formula:

$$\text{ET (cm/month)} = 16\left(\frac{10T}{I}\right)^a \tag{1}$$

where T is the monthly temperature (°C) and I is the annual heat index defined as

$$\mathrm{I} = \sum_{i=1}^{12} (Ti/5)^{1.514} \quad \mathrm{i: 1{\sim}12} \tag{2}$$

$$a = (492390 + 17920\,\mathrm{I} - 77.1\,\mathrm{I}^2 + 0.675\,\mathrm{I}^3) \times 10^{-6}$$

Considering a maximum temperature of 26.5 °C, the calculated evapotranspiration was 774 mm/yr in Ogaki, 767 in Kuwana, and 747 in the Aichi station. These values indicate that 41 to 48 % of the rainfall is lost to the atmosphere.

Table 1. Physical characteristics of the geological system

Formation		Main lithology	Hydraulic conductivity (cm/sec)	Thickness* (m)
Nanyo		Fine sands - silt -	3.2×10^{-3} to 9.9×10^{-8}	36.4
Nobi		clay	4.2×10^{-4} to 7.8×10^{-8}	20.7
First Gravel Bed		Clays - silt	2.1×10^{-3} to 1.1×10^{-4}	15.8
		Gravel - sands		
	upper	Fine sands	1.9×10^{-3} to 6.9×10^{-7}	60
Atsuta	lower	Sands - clays	2.1×10^{-3} to 4.4×10^{-8}	42.1
Second Gravel Bed		Gravel - sands	2×10^{-3} to 3.3×10^{-3}	25

* measured at Aburashima.

3. Methods of Study

Groundwater modeling is a powerful management tool which can serve multiple purposes such as allowing quantitative prediction of the responses of the hydrologic system to externally applied stresses (Senthilkumar and Elango, 2004). Thus, after gathering the hydrogeological data of the area a three-dimensional model was constructed to understand the groundwater flow in the Nobi Plain. The model was refined around Aburashima to account for local details. Based on the flow simulation, a series of scenarios were analyzed to estimate the maximum amount of groundwater that can be extracted at the site of interest without leading to land subsidence.

The investigation started by drilling three exploratory wells to a maximum depth of 200 m. These wells permitted define the geometry and physical properties of the sediments in the site, which constituted a basic input to the model.

The next phase of the research was based on numerical modeling using the modular three-dimensional finite-difference code Modflow (McDonald and Harbaugh, 1988). This program is widely applied for simulating flow in complicated basins with various natural and/or artificial hydrogeological processes (Ebraheem et al., 2004). Modeling of the groundwater flow through porous media is performed by solving the governing equation:

$$\frac{\partial}{\partial x}\left(Kxx\frac{\partial h}{\partial x}\right) + \frac{\partial}{\partial y}\left(Kyy\frac{\partial h}{\partial y}\right) + \frac{\partial}{\partial z}\left(Kzz\frac{\partial h}{\partial z}\right) - W = \mathrm{Ss}\frac{h}{\partial t} \tag{3}$$

where *Kxx*, *Kyy* and *Kzz* represents the saturated hydraulic conductivity along the x, y and z directions; *h* is the potentiometric head above a common datum; W is a volumetric flux per unit volume and represents sources and/or sinks of water; *Ss* is the specific storage of the porous medium, and *t* corresponds to time. Since the simulation was run under steady state, the terms on the right were neglected. The model domain was replaced by a series of discrete points in space, and hydraulic heads at each one of them was obtained solving the governing equation. After development, the flow model was calibrated to the site conditions to ensure it provided a reasonable representation of the real situation. On the basis of the computed results, the simulation was finally used to determine the safe yield of the aquifers at Aburashima.

4. Numerical Modeling

4.1. Concept and Data Input

The model domain was delimited by the Yoro Mountains and Owari Hills to the west, north, and northeast, the Ise Bay to the south, and the Shounai River to the southeast. Thus, a large part of the region was circumscribed by impermeable boundaries, while a constant head of 0 m was assigned to the bay. River cells were imposed along the major bodies of surface water. For management purposes the city of Nagoya constitutes an independent entity therefore, it had to be excluded from the analysis. The geology of the area was divided into seven layers, two of them representing the upper and lower aquifer, the rest attributed to the different confining units. Each layer was assumed to be continuous, with a general tilt westwards. The surface elevation was set in accordance with the digital maps edited by the Geographical Survey of Japan (1997).

The model grid totalized 314 rows and 198 columns. Cells were of variable size although they were refined to a maximum of 20 m by side around Aburashima to get better representation of the piezometric contours (Figure 3). Hydraulic conductivity was assigned to each individual layer, with values ranging from 2 x 10^{-3} in the sands and gravels, to 9 x 10^{-8} cm/sec in the clay horizons.

Aside from rainfall, recharge to the aquifers is dependant on variables such as land use, soil type and moisture, evapotranspiration rates, irrigation, topography, and period of the year. A detailed evaluation of all these contributions is highly complex so, the overall recharge to the upper layer was initially approached from the difference between rainfall and evapotranspiration, and then reasonably adjusted during the calibration process to reproduce the observed site-conditions.

Pumping was represented by negative fluxes which incorporated the groundwater withdrawal in 822 wells at the First Gravel Bed, and 737 wells at the lower aquifer. Pumping wells in Gifu prefecture were grouped in coincidence with the main industrial centers, while for Aichi, they were assigned in accordance with a 1 km x 1 km mesh maintained by the prefecture. Extraction rates were represented as the annual mean discharge for each well during 2004, coincident with the most recent information available.

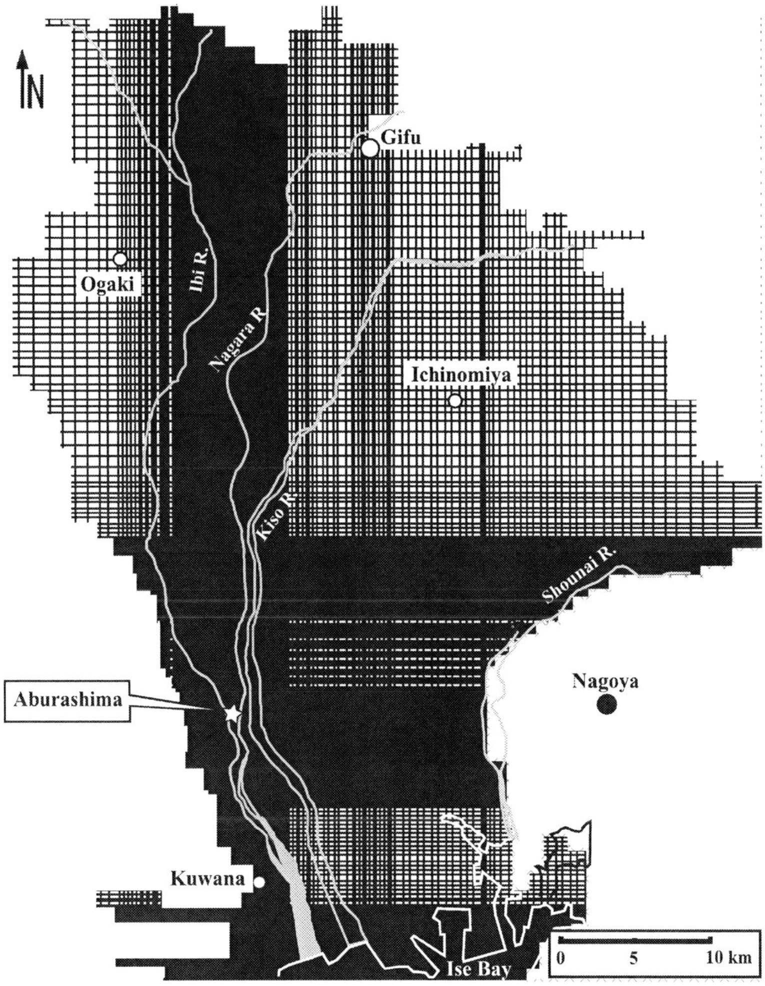

Figure 3. Model domain and finite-difference grid.

4.2. Model Calibration

The steady-state model was calibrated against in-situ data until simulated results reasonably matched the conditions observed in a network of 69 monitoring wells, providing a good coverage of the Plain. In agreement with the groundwater abstraction data, calibration was based on potentiometric observations for 2004. The procedure consisted in systematically varying the poorly known values of recharge and river stages, without introducing any changes to better defined parameters. When results were not satisfactory, the input values were modified and a new calibration process was started. In a second step, the optimization process was restricted specifically to the site of interest until no significant improvements in the results were obtained. The best outcomes were achieved after dividing the model domain in 11 regions of recharge with rates from 0 to 900 mm/yr. Maximum recharge would take place on the northwest along the boundary with the Yoro Mountains, while values can be considered minimum towards the lowlands in the center and south. Water stage and the conductance of rivers and streams were also adjusted to produce more realistic results. Figure 4 shows the difference between monitored and computed heads after calibration. The mean residual for the model is 0.048 m, which represents 0.15 % of the total head difference in the entire region (19.3 to -12.8 m). The absolute residual reached a minimum of 0.012 at the Tsushima monitoring station, but on the contrary, major bias led to poor results at Ichinomiya and surroundings, with a difference in heads up to 4.7 m. A misrepresentation of the realistic situation in that zone may be attributed to an insufficient grid refinement. In effect, the elevation of the water table varies drastically over short distances resulting thus in an especially high hydraulic gradient which can barely be represented by the selected mesh size, approximately 500 m by side in the area. Unfortunately, a further mesh refinement was not possible as it resulted in numerical instability and therefore, limited the capacity of the model to accurately simulate the groundwater heads at specific locations.

Statistics parameters were also used to evaluate the accuracy of the calibration. The normalized root mean squared is the root mean squared error divided by the maximum difference in the observed head values. A reasonable calibration can be expected when it is below 10 % (Waterloo Hydrogeologic, 2005). The present simulation yielded a value of 6.7 %. In addition, a histogram of the residuals indicates that most of the residuals followed a normal distribution with the mean value near zero. As stated above, deviations in the distributions corresponded to the Ichinomiya surroundings (Figure 5).

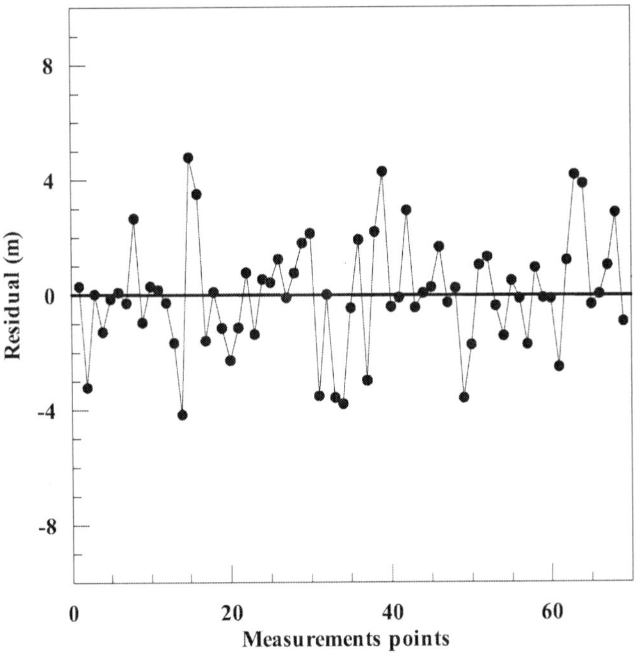

Figure 4. Residual of computed and observed heads at the monitoring wells.

Finally, an overall mass balance of the region yielded negligible differences between inflows and outflows constituting thus another indicator of a successful simulation. In this regard, approximately 48 % of the total water input came from infiltration through the riverbeds, while rainfall contributed with 37 % of the recharge and the constant heads cells 15 %. In contrast, the outputs corresponded to wells abstraction (52 %), discharge through constant heads (24 %), flow to rivers (16 %), and a minor part to evapotranspiration (8 %). A sensitivity analysis was also carried out to test the response of the model to variations in the key input parameters. It was found that the model is especially sensitive to changes in river conditions, as simulated heads for the upper aquifer decreased linearly with a reduction in the river stages. These results emphasize that river information must be carefully managed, as an error in the estimations translates into deviations of the same magnitude in the heads at the site of interest.

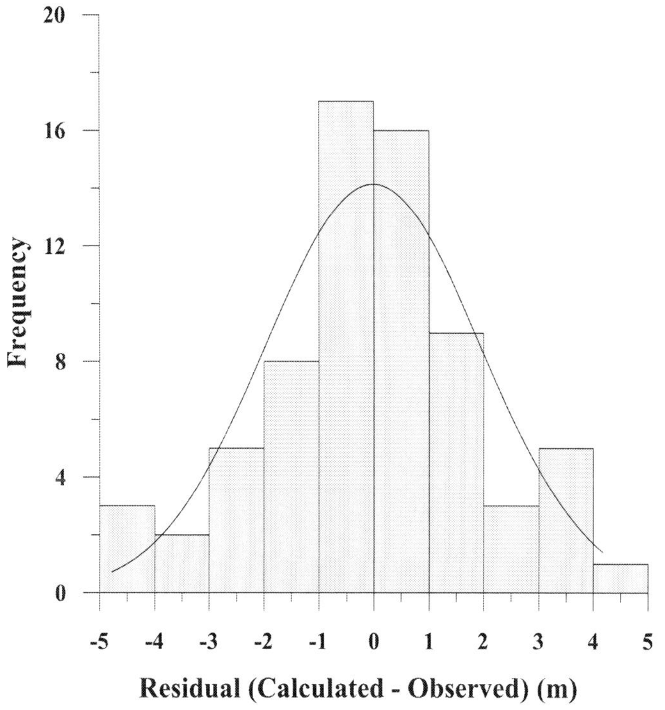

Figure 5. Distribution of residuals throughout the Nobi Plain.

4.3. Flow Conditions

Head contours indicate that groundwater flows mostly from northwest to southeast. There is an additional component that originates at the foot of the Yoro Mountains and moves from west to east. Groundwater flow converges from all sides towards the center of the plain discharging ultimately through the Shounai River (Figure 6). There are many pumping wells used for industry in that part of the plain, which causes a low hydraulic head (Uchida et al., 2003). This statement was confirmed by the negative hydraulic heads simulated all over the central region.

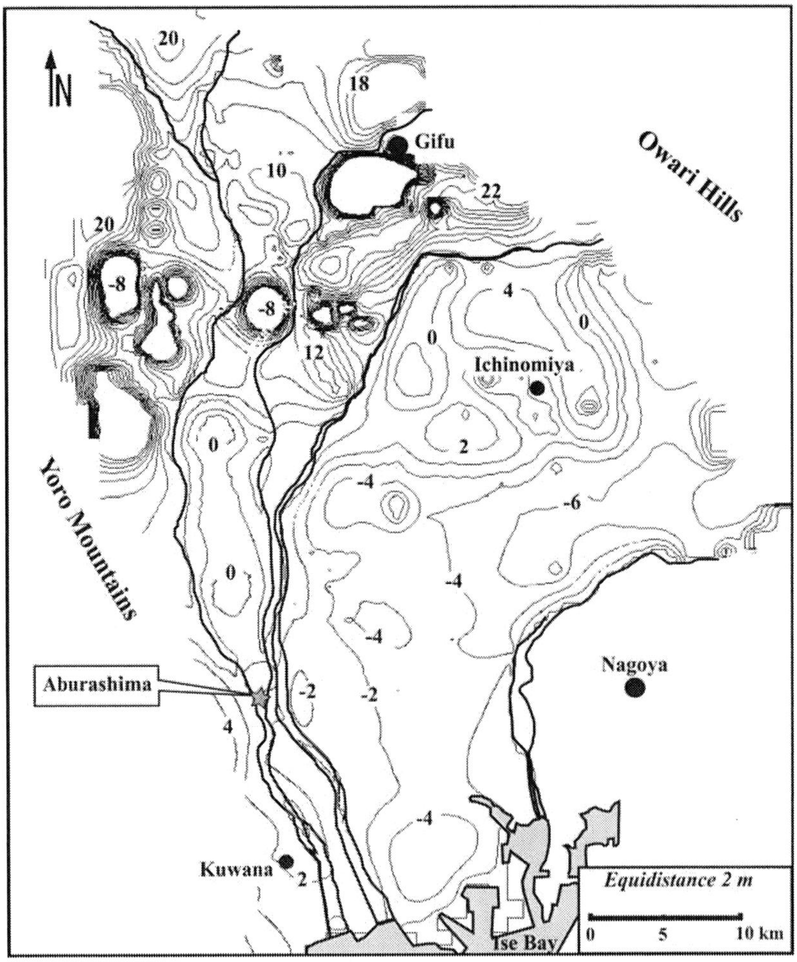

Figure 6. Simulated hydraulic heads for the shallow aquifer.

It is widely known that prolonged overdraft of groundwater can lead to salt contamination of freshwater aquifers. According to the simulation, groundwater abstraction resulted in the advance of seawater to a distance of about 2.5 km inland through the shallow aquifer, defining a mixing zone that encompasses the towns of Kuwana, Matsunaka, Yatomi and Tobishima near the coast. Exploitation of the deep aquifer is more limited and consequently, inland penetration of the saline wedge seems not to occur as a homogeneous front, but rather as a reduced tongue restricted to the surroundings of Yatomi and Tobishima. Chemical samples of the aquifers revealed that waters rich in Na-K and Cl^- dominate throughout the south of the plain spreading several km inland from the shoreline (Figure 7). Chloride concentrations as high as 3.5 g/L have been occasionally measured in the groundwater of Aburashima (METI, 2007). Unlike the sharp transition zone between fresh and brackish groundwater near the coastline depicted by most hydrogeology textbooks, saline groundwater was found far inland as occur in other coastal plains like in The Netherlands, Suriname, and Java (Groen et al., 2000). Although groundwater exploitation is important, the current extent of salinity should be interpreted carefully because it can be influenced by natural processes in addition to seawater intrusion. Chloride concentrations seems to increase together with a decrease in the sediments grain-size (METI, 2007), suggesting a possible contribution of fossil waters flushed from the marine clays and silts intercalated within the confining units. This hypothesis must be tested by further geochemical surveys supported by isotopic tracers, which constitute one of the most useful tools to discriminate Na-Cl sources in groundwater. Regardless the salinity origin, the hydrochemical zonation of the aquifers is a critical aspect to consider when planning future developments in the area, as groundwater of poor quality may be found in the new boreholes. Thus, utilization of the resource may become problematic for certain uses as agriculture, since high salinity of waters severely affects plants growth and reduces yields.

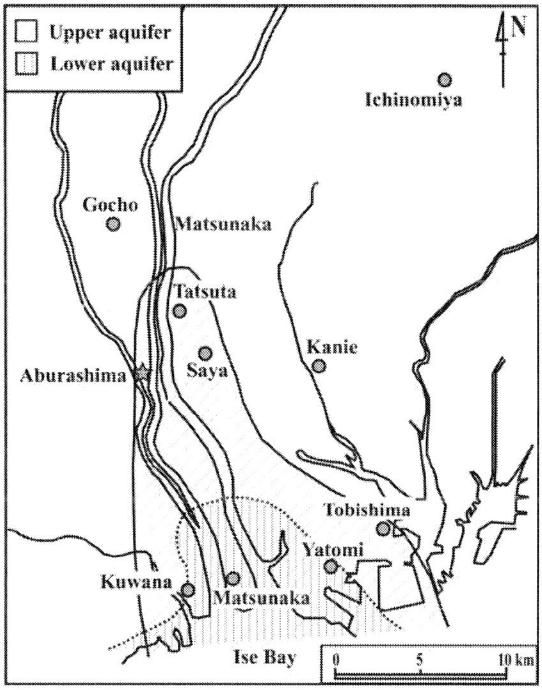

Figure 7. Spatial extent of saline waters (after METI, 2006).

5. Predictions for Aquifers Exploitation

5.1. Groundwater Availability

The validated model was ultimately used to calculate the availability of groundwater with the aim at maximizing the pumping extraction from the confined aquifers while preventing subsidence. Long-term monitoring at Gōcho station showed the site is especially susceptible to land subsidence when the groundwater of the upper aquifer drops more than 0.8 m below the average level (METI, 2007). Given its proximity, it is expected the system at Aburashima will respond in a similar manner to head fluctuations. Since there was a difference of 0.17 m between model results and observations, predictions at Aburashima conservatively considered the permissible groundwater drawdown cannot exceed 0.63 m. In contrast, data at the Tsushima and Saya stations showed that during 1985-1986, a reduction in pumping from the

Second aquifer was accompanied by a rebound of groundwater levels above -10 m, giving place at the same time to the stabilization of subsidence rates to a minimum. From the above information, it can be concluded that sustainable pumping in the Aburashima site can only be achieved if the maximum drawdown does not exceed 0.63 m for the upper aquifer, and 10 m for the deep unit.

It was calculated that the maximum supply capacity of one well at the shallow aquifer is 27.7 m^3/day, while the deep one permits to extract 28 times more water, with a maximum rate of 776.4 m^3/day. A possible strategy to increase the extraction volumes consists in placing a number of wells in a circle centered in Aburashima. As can be anticipated, raising the number of wells is corresponded with an increase in the pumping rates (Figure 8). However, the relation is not nonlinear, especially for the deep aquifer, which means the extraction efficiency reduces for larger rates. Generally speaking, constrains in the exploitation efficiency translate directly into a higher cost for water supply. Then, it is possible to estimate the best well design from the economical point of view. Considering the site location, sediments characteristics, and the current market conditions in Japan, the cost for constructing an extraction well to the upper aquifer (60 m) is about U$$ 40,000. The price approximately duplicates when drilling 175 m to the Second Gravel Bed (Hytec Co., personal communication). Then, it was found that for the shallow aquifer, the best compromise between exploitation rates and cost is attained for 5 wells, where the price of water reaches a minimum (Table 2). As shown, at the early stages the price is about 44.5 $/$m^3$, but it decreases exponentially to 3.7 $/$m^3$ after 1 year of pumping, as the investment is better amortized over longer consumption periods. The cost of water delivery is also a function of the vertical distance, as deeper aquifers require a lift over greater distances and against a higher gravity force. In the study site, the benefit of extracting much larger volumes of groundwater from the lower aquifer overcomes the higher costs of lifting, resulting in a more efficient relationship extraction-price.

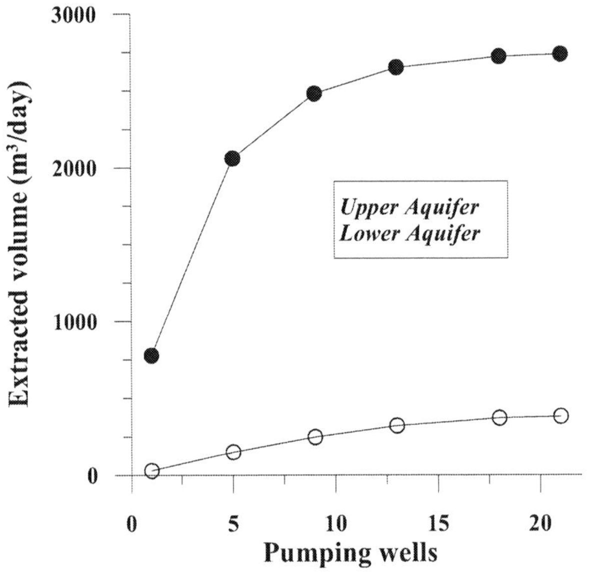

Figure 8. Maximum pumping rates in relation to the number of wells.

Table 2. Cost of groundwater extraction as a function of the number of wells and period of pumping. Unit: U$/m^3

No. wells	G-1				G-2			
	30days	**90days**	**180days**	**1year**	**30days**	**90days**	**180days**	**1year**
1	48.1	16.0	8.0	4.0	3.4	1.1	0.6	0.4
5	44.5	14.8	7.4	3.7	6.5	2.2	1.1	0.7
9	48.6	16.2	8.1	4.0	9.7	3.2	1.6	1.1
13	54.2	18.1	9.0	4.5	13.1	4.4	2.2	1.5
18	65.0	21.7	10.8	5.3	17.7	5.9	2.9	2.0
21	73.5	24.5	12.3	6.0	20.5	6.8	3.4	2.3

As an example, a $ 720,000 investment will allow the construction of 18 wells into the first aquifer and only 9 into the deep one however, it is more effective to extract water from the lower unit as there are both a 6.7 times increase in production and a simultaneous drop in the prices with respect to the shallow aquifer. It must be emphasized that the analysis above is far from being a study of costs but instead, it must be treated as a complementary exercise to help finding an optimal solution in terms of total pumping rate and consumer benefits. The water price was calculated on the cost of wells and pumping volumes alone therefore, it reflects the bottom-end value expected as other factors like wells maintenance, energy requirements, and operating conditions will add up to the true extraction expenses. Furthermore, the prices are sensitive to the period of production, decreasing exponentially especially after 6 months of withdrawal.

5.2. Safe Yield

Calculations were done under steady state to determine the mean groundwater available for exploitation. Nevertheless, intensive withdrawal for irrigation of rice fields in summer causes some temporal declines in the groundwater levels that reduce the supply capacity of the aquifers. Data collected during the second half of 2006 indicate that groundwater heads in Aburashima declined up to 0.24 m respect the values used for the model predictions. This drop in heads is negligible for the lower aquifer with an allowable drawdown of 10 m however, it becomes highly significant for the shallow unit, where the permissible drawdown is in the order of cm. Low groundwater-levels may last from a few days to several months therefore, the extraction policies for the First Gravel Bed should be defined on a seasonal basis, otherwise determined in relation to the deepest heads observed in the field. With a water table falling to a minimum, the permissible drawdown of the upper aquifer reduces and in consequence, the extractable flow rates decreases between 38.5 and 44.5 %, nearly 8.5 to 12 m^3/day per production well. These results show the strong influence that agricultural activities exert on the supply capacity of the shallow aquifer so, they must be carefully considered in the establishment of new abstraction policies. Furthermore, changes in groundwater recharge due to climatic variations as droughts, fluctuations of the stream flow in rivers, and the effect of local heterogeneities may result in incorrect determinations of the limit of pumpage from the aquifer. Thus, an effective management of the water resources must consider all components of the hydrologic system to maintain a sustainable yield in the long term (Sophocleus, 1988; Sakiyan and Yazicigil, 2004). Therefore, a "safety factor" of 20 % was added to the supply capacity of the aquifer when heads are at their lowest, so as to ensure that pumping will not lead to subsidence even during periods of reduced groundwater availability. These rates constitute the safe yield of the aquifer, and can be regarded as the maximum amount of groundwater available for stakeholders in Aburashima (Table 3).

Table 3. Optimal rates of groundwater extraction from the First and Second Gravel Bed

Aquifer	Number of wells	Maximum withdrawal (m^3/day)	Safe yield (m^3/day)
	1	15.4	12
G-1	5	91.7	73
	9	151.8	121
	1	776	621
G-2	5	2059	1647
	9	2479	1983

Conclusion

Land subsidence due to groundwater overdraft is an environmental issue that has been affecting the Nobi Plain for several decades. Strict controls on water usage brought the problem under control however, the spread of new economical activities especially towards urban outskirts indicate that the current regulations will soon be insufficient to meet the users needs. The present work presented a numerical simulation that enhanced the understanding of the flow patterns, and predicted the safe yield of the First and Second aquifers for a sustainable development of the Aburashima region. In that case, the objective was to maximize the groundwater withdrawal while avoiding negative consequences. Results showed that water availability in the upper aquifer is somewhat restricted and therefore, abstraction from the lower unit would be more effective in terms of supply and relative cost. Moreover, seasonal variations, and a higher demand for irrigation during summer have a significant impact on the exploitation of shallow wells, but are irrelevant for deep aquifers. When required, a feasible strategy to increase the abstraction rates is to distribute pumping wells in a network around the target site so as to cover larger areas of the aquifers.

It is anticipated that management authorities will utilize the present work in their efforts to adequate a sustainable supply to the present and future demand of groundwater. Although effective, restrictions on groundwater abstraction are only part of the solution since the continued growth of industrial and agricultural developments soon or later will derive in new conflicts between withdrawal and the ability to provide the resource. Control of groundwater depletion is not only a hydrogeologic problem, but it also involves political, socioeconomic, and technological factors. Therefore, a key point is to search for an integrated plan to manage the natural resources in the plain, which indefectibly will require the involvement of a number of governmental agencies and private parties. The cooperative work might be able to explore alternative measures to reduce the groundwater dependence, which indirectly would mitigate the subsidence threat. For example, it might be possible to evaluate the technical and economical feasibility of diverting surface waters for irrigation, construction of ponds and reservoirs for storage, increase in energy tariffs or taxes for larger groundwater consumers, restrict future industrial wells to deep aquifers, and implement subsides for cultivation of crops with low-water requirements.

Acknowledgements

The authors gratefully acknowledge Dr. Frank Columbus for the invitation to prepare the present paper.

References

Don, NC., Araki, H., Yamanishi, H., and Koga, K (2005). Simulation of groundwater flow and environmental effects resulting from pumping. *Environ. Geol.* **47**, 361-374.

Ebraheem, AM., Riad, S., Wycisk, P., and Sefelnasr, AM (2004). A local-scale groundwater flow model for groundwater resources management in Dakhla Oasis, SW Egypt. *Hydrogeology J.* **12** (6), 714-722.

Fetter, CW (2000). *Applied Hydrogeology.* 4th Edition. Prentice-Hall Publish, USA Geographical Survey of Japan (1997). *Digital Map 50 m Grid* (Elevation), Japan-II.

Groen, J., Velstra, A., and Meesters, AGCA (2000). Salinization processes in paleowaters in coastal sediments of Suriname: evidence from δ^{37} Cl analysis and diffusion modeling. *J. Hydrology* **234**, 1-20.

Japan Meteorological Agency (2007). Climatic Report of the Japan Meteorological Agency, available from *http://www.jma.go.jp/jma/indexe.html.*

Larson, KJ., Başağaoğlu, H., and Mariño, MA (2001). Prediction of optimal safe ground water yield and land subsidence in the Los Baños-Kettleman City area, California, using a calibrated numerical simulation model. *J. Hydrology* **242**, 79-102.

McDonald, MG and Harbaugh, AW (1988). A modular three-dimensional finite-difference ground-water flow model. US Geological Survey Techniques of Water Resources Investigations, Book 6.

Ministry of Economy, Trade and Industry of Japan (METI) (2006). Survey on the rational use of groundwater in the city of Kaizu, Gifu prefecture. *Report to the Bureau of the Ministry of Economy in Central Japan,* **117** pp (in Japanese).

Ministry of Economy, Trade and Industry of Japan (METI), (2007). Survey on the rational use of groundwater in the city of Kaizu, Gifu prefecture. *Report to the Bureau of the Ministry of Economy in Central Japan,* **73** pp (in Japanese).

Sakamoto, T., Takada, Y., Kuwahara, T. and Itoigawa, J (1986). Geology of the Nagoya-nambu district. Quadrangle series, scale 1:50,000, Kyoto (11) No. 32. *Geological Survey of Japan.* (in Japanese with English abstract).

Sakiyan, J. and Yazicigil, H (2004). Sustainable development and management of an aquifer system in western Turkey. *Hydrogeology J.* **12**, 66-80.

Senthilkumar, M. and Elango, L (2004). Three-dimensional mathematical model to simulate groundwater flow in the lower Palar River basin, southern India. *Hydrogeology J.*, **12**, 197-208.

Sophocleus, M (1998). Perspectives on sustainable development of water resources in Kansas, Bulletin 239, *Kansas Geological Survey*, Lawrence, Kansas, USA, pp 239.

Thornthwaite, CW. (1948). An approach toward a rational classification of climate. *Geol. Rev.* **35**, 55-94.

Uchida, Y., Sakura, Y., and Taniguchi, M (2003). Shallow subsurface thermal regimes in major plains in Japan with reference to recent surface warming. *Phys. Chem. Earth* **28**, 457-466.

Yamamoto, S. (1984). Case History No. 9.6. Nobi Plain, Japan. In J. F. Poland (Ed.), *Guidebook to studies of land subsidence due to ground-water withdrawal*, pp 195-204, UNESCO International Hydrological Programme, Working Group 8.4, USA.

Yamanaka, M. and Kumagai, Y (2006). Sulfur isotope constraint on the provenance of salinity in a confined aquifer system on the southwestern Nobi Plain, central Japan. *J. Hydrology* **325**, 35-55.

Yoshida, F., Kurimoto, C., and Miyamura, M (1991). Geology of the Kuwana district. Quadrangle series, scale 1:50,000, Kyoto (11) No. 31. *Geological Survey of Japan*. (in Japanese with English abstract).

Waterloo Hydrogeologic (2005). *Visual MODFLOW* v. 4.1 User`s Manual. Waterloo, Ontario.

In: Focus on Urbanization Trends
Editor: Laura M. De Smet
ISBN: 978-1-60456-109-8

Chapter 7

Spatial Effects of the Interaction between Environment and Urban Area and Approaches to Coordination in Xi'an

Dongqian Xue
College of Tourism and Environment,
Shaanxi Normal University, 99 Changan Road, Xi'an, China

Abstract

Through analyzing the interaction between urbanization and environment, we sum up the spatial morphology, formation mechanism and types of the interaction between urbanization and environment. Then we generalize the spatial effect types of the interaction, namely tree-ring effect, decay effect and encroachment effect. In the case study of Xi'an City, we find out the spatial effects of the interaction between urbanization and environment and their spatial representation styles. Finally, we point out the macroscopic, mesoscopic and microscopic coordination arrangements, and our thoughts on the coordination in development mechanism, development structure, development ways and development goals.

Keywords: spatial effects, spatial morphology, spatial coordination, Xi'an.

1. Theories on Interactions between Urban Area and Environment

1.1. Spatial Evolutions

One of the two aspects of spatial evolution is the spatial expansion, and the other is the increase of spatial dispersion and the decrease of spatial concentration. As to the spatial expansion, land use scale expands with the expansion of population in urbanization, and there is spatial extensiveness and peculiarity in the interaction between urban elements and environmental elements, such as the partial enlargement of noise pollution, the concentration of water pollution, and the formation of ground subsidence, etc. On the other hand, together with the spatial expansion of urban area, elements in urban system trends to disperse when traffic and communication system develops, which will decrease the elements' requirement to urban space. Urban spatial allocation and the types of interaction also trend to disperse, and urban structure differentiates spatially in some cases. The concentration and dispersion of urban elements act simultaneously in urbanization. If concentration trend is greater than dispersion trend, city and the interaction types show a status of concentration, whereas the status is dispersion [1].

1.2. Spatial Morphology

The spatial morphology of interaction between urban area and environment is mainly determined by the distribution of pollution sources, dispersion path and pattern of pollutants, etc. Urban industrial allocation, which can be classified into such three types as concentrated, dispersed and randomly allocated industry, is crucial to the spatial morphology of this interaction between urban area and environment because industry is the biggest pollution source.

Concentrated industry relies greatly on central business district, infrastructures and resources distribution, because it must obtain raw material from ambient area or regional accessory firms, and the products should be forward or backward linked, and the cooperation should resemble or be mutual complementing to other firms in this area. Dispersed industry relies less on central business district or other firms. The first reason is that it obtains raw materials from other regions and requires not so much to such conditions as regional water resource. Secondly, its market is not confined to local area. Thirdly, dispersed industry has a larger economic scale and occupies more land than concentrated industry, and its firms are in a relatively integrated system. Fourthly, dispersed industry relies greatly on long distance transportation. Randomly allocated industry, whose characteristics are high-fine processed or assembled products, technology-intensive or labor-intensive industries, obtains ubiquitous raw materials and resources, and its market includes the urban area and its ambient area [2].

The spatial morphology of the interaction between urban area and environment, which is constructed by the allocation styles of industries and city, has three types, linear, multiple-nuclei and sector spatial morphology. Linear spatial morphology is formed by the axis of transportation line, river or stripe topography, which causes the linear morphology of the interaction processes and results. The case is water pollution in steams. Multiple-nuclei morphology is spatially an industrial region based on large-scale firms or residential region, showing special in spot-like environment, such as local ground subsidence, local air pollution exceeding standard, etc. Sector morphology is spatially a large-scale comprehensive industrial region, where all kinds of pollution and environmental disasters spread from the central pollution source and the pollution degree decreases gradually.

The interaction between urban area and environment occurs in many spatial levels. Air pollution, water pollution and solid wastes pollution, which spread from industrial region to business and residential region, then to others, can be monitored in industrial region, the main polluted urban area. The main pollution in business region is noise pollution, solid garbage and dust, related to urban population density, urban environment and spatial location to industrial region. Comprehensive pollution occurs in other functional regions with small scale and low intensity. But, when pollution intensity in ambient urban area is greater than that in urban area, pollution spread from central urban area to urban fringe zone, then to ambient urban area, or spreads inversely when pollution intensity is greater in urban area. This is determined by the spatial location of pollution resource and its type and intensity. And this phenomenon also occurs when pollution is noise pollution or daily life pollution. Vertically, pollution spreads from one-dimension urban area to two-dimension urban area, then to three-dimension urban area. Analyzed from the angle of spatial elements, pollution spreads from points to lines, then to area. Points here refer to firm, industrial region, urban population-dense area, main resource points, etc. Lines here refer to transportation lines, water transportation lines, stripe topography, airflow channel, and so on. Area here refers to each independent regional unit, functional area and urban area, etc.

The pollution's spatial morphology in urban area spreads in three ways, circular-layer pattern, linear pattern, and hop-and-leap pattern. In circular layer pattern, pollution spreads continuously from inner source, like the distribution of tree ring. Polluted area expands gradually with relatively low speed, but the influence declines simultaneously. It is difficult to control this kind of pollution, such as water and soil erosion, pollution in ground water and soil. When pollution spreads in linear pattern, pollutant disperses along the axis because of low friction in the axis. Then, a relationship is formed between the pollution source and the newly polluted area, while the axis area is the mostly polluted. Accumulative effect and low friction in the axis, especially the distribution of many firms along the axis, accumulate the pollutant and aggravate the pollution. In difference to circular-layer pattern, pollution degree in linear pattern is higher in lower reach than in upper reach, for instance, the pollution in rivers and noise pollution. Hop-and-leap pattern is not a continuous spreading pattern. Hop-and-leap pattern appears when continuous pattern breaks off, because of characters of geo-environment or pollutant carrier, or sometimes because of policies and planning, which can be seen in the case of noise pollution, air pollution and ground subsidence, etc.

The spatial morphology of this interaction is constrained by many kinds of factors. Firstly, natural factors, such as urban geology, topography, surface water and ground water, climate, vegetation, are the fundamental condition to have influence on the structure of morphology after their geographical location being determined. Secondly, urban function determines possible environmental problems and characters of interaction, thus affects urban scale and structure. Cities in different scales have different spatial morphology, the bigger the more complicated. Thirdly, urban planning and policies on urban development, environment, resources, and related laws are usually used in urban management and control. In the past, urban development depended mainly on investment and its related policies in China, which played an important role in morphology evolution in different period or stage. Investment and its related policies can prescript urban space and the pollution space, and control all kinds of environmental pollution or damage. Fourthly, socio-cultural factors, including social structure, life style, residents' social behavior criterion and cultural value conception, etc. affect urban environment and its morphology, while environment change and its spatial morphology affect human's socio-cultural psychological activities [3].

1.3. Spatial Effects

Spatial effects of the interaction are the changes of economic density, pollutant density, social factors' density and the condition of systems in interaction under the disturbances outside, for instance, acid rain caused by change of pollutant density. Spatial effects can also be spatial lagging effects, that is, spatial effects of the interaction are far away from effect source. The case is that acid rain area usually appears far away from pollution source, or pollutant accumulates in the lower reach of rivers, etc. Spatial effects can be shown in such three forms as tree-ring effect, attenuating effect and encroaching effect.

1.3.1. Tree-Ring Effect

Urban economy is the fundamental motivation of land use and its spatial structure evolution in the fringe of urban-rural ecotone. The periodical rule of economy development results in the rhythm change of land use and cover change (LUCC) in the fringe of urban-rural ecotone, which can be categorized into accelerating period, stable period and retarding period. The expanding speed of urban non-agricultural land and the general characteristics of LUCC are various in different conditions, similar to the increase of tree ring.

Firstly, periodical economy development results in periodical changes of expanding speed of urban land use. Urban land use, transportation land use and residential land use will increase rapidly and the conversion speed from agricultural land to non-agricultural land accelerates obviously, because of the production scale expands and the revenue and investment increase during rapid economy development. But when economy is in its stagnant or declining stage, the expanding speed of urban land use will slow down greatly. Secondly, periodical economy development affects the exploitation degree and benefit of agricultural land. Accelerating economy development means higher level of people's daily life and larger production scale, and agricultural land continually converts into urban area, all of which require intensive management of agricultural land and higher land use level. Simultaneously, economy development is the fundamental condition of intensive management of agricultural land [4].

1.3.2. Decay Effect

Distance decay effect has a spatial implication here, referring to the decay phenomenon of influence of natural elements and socio-economical objects in regional space when distance among them increases, which can be illustrated by the following formula.

$$I = f(\frac{1}{d})$$

In this formula, I refers to influence, and d refers to spatial distance, and f refers to friction coefficient. As is the cases, the aggregation or dispersion effect of urban economy, or environmental effect in exploiting natural resources, or the pollution effect of urban pollutant source, will decline gradually if distance increases from city to its surrounding area [5].

1.3.3. Encroachment Effect

In regards to primary environmental system, the distribution of resources, regional land use and environment quality are stable and harmony to each other. But, the former integrated and stable system structure was affected by many factors in the accelerating process of urbanization, industrialization and modernization. The system structure missed spatially and distributed irregularly, and the change of former structure is Encroachment Effect, such as the changes of land use types, land use structure and its spatial scale, or local polluted area in a clean environment, or destruction to the environmental integrity resulted from exploiting resources, etc.

2. Spatial Effects of the Interaction between Urban Area and Environment

2.1. Spatial Effects of Water and Soil Erosion

Exploitative water and soil erosion is resulted from extravagant ground disturbance in urban land exploitation or infrastructure construction, and the erosive modulus is not less than 100,000 km^3/a. Ultra-speed urban development and large-scale construction caused channel deposition and drainage deposition, letting down the city flood control ability and destructing the city infrastructure, and seriously threatening urban environment quality and sustainable development of urban economy [6]. For example, most of the water and soil erosion in Xi'an city occurred in urban fringe zone, all kinds of development zone, stone-quarry and soil-quarry area, site of large engineering project, etc.

City can be spatially divided into urban built-up area, suburb and outer suburb, regionally a transitional relationship. There are many differences among them in land use pattern, land cover type, natural or social environment and process, mechanism of the interaction between human and the earth, etc. Therefore, the density of water and soil erosion is of great difference in their characters, which can be illustrated by Figure 1.

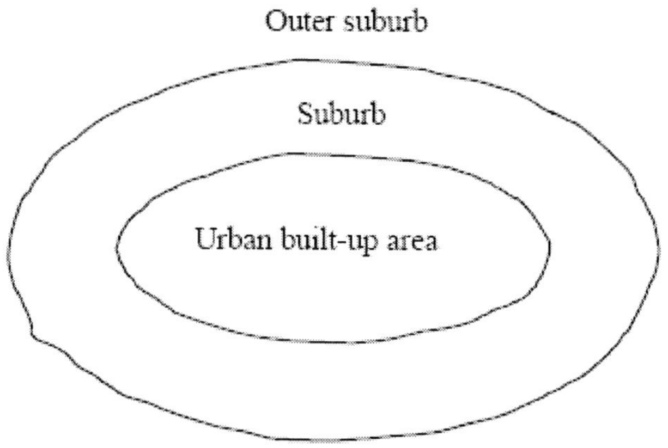

Water resource loss: outer suburb < suburb < urban built-up area
Soil resource loss: urban built-up area < outer suburb < suburb

Figure 1. Sketch map of regional difference in urban water and soil erosion area (according to Yuan Renmao).

2.1.1. Urban Built-Up Area

Urban built-up area is typically an artificial environment, where there is mass impermeable earth surface and well-arranged drainage system. Because of the large proportion of impermeable ground, usually greater than 75 percent, the permeability coefficient of surface water in urban area is lower than that in suburb and far outskirt, and surface water flows out rapidly in the well-arranged drainage after aggregated. Though the amount and speed of surface water increased in the built-up area and thus enhanced the erosion ability, the ground erosive modulus is far lower than suburb area and rural area, usually 20 to 80 t/km2·a under the protection of such earth surface as cements, etc. Simultaneously, low permeability coefficient in urban area caused great water resource loss and low ground water replenishment.

2.1.2. Suburb: Urban-Rural Ectone

It is the ectone from urban area to rural area, where environment changes rapidly because of industrial or agricultural production and land exploiting activities. In this zone, many kinds of highly intensive activities also caused serious water and soil erosion. Construction and exploiting activities in suburb changed the land properties and landform, and destroyed the former balance between environment and water and soil erosion. To reach a new balance is to change the intensity characters of water and soil erosion. Because of disturbance from humankind, the anti-erosion ability of surface soil weakened obviously, which relatively increased the erosion ability, and the speed and intensity of water and soil erosion are far greater than those in built-up area and outer suburb [7]. In some development zone in Shenzhen, the greatest erosive modulus is as high as 131 kt/km2·a, and the average erosive modulus is 61.1 kt/km2·a [8]. This zone is the most seriously affected by water and soil erosion, thus we call it Water and Soil Erosion Belt.

2.1.3. Outer Suburb

Except the human effect in farmland area, natural effects in water and soil erosion prevail in all other area. Therefore, soil erosion in rural area varies greatly according to natural or human factors, and the erosive modulus usually is 10 to 15 kt/km2·a. Because of high ground permeability coefficient and better vegetation cover there, which means strong water conservation ability, the surface water feeding amount to ground water is fairly greater than urban or suburban area, while the average water and soil erosion amount is lower.

In general, the distribution of water and soil erosion in urban area is similar to multiple-nuclei or sector morphology type in the spatial morphology of the interaction between urbanization and environment, expanding in circular-layer pattern, which caused the tree-ring effect spatially.

2.2. Spatial Effects of River Pollution

2.2.1. Status of River Pollution

The condition of river pollution can be illustrated by the comprehensive pollution index. Seen from its general trend, the pollution condition of Wei River and Zao River trends to drop annually, while the others rise slightly or drop little. As to the pollution degree, Wei River and Zao River are the two most seriously polluted, Chan River and Feng River followed, while the two most lightly polluted rivers are Ba River and Jue River (Figure 2). Pollution degree is mainly determined by pollutants' spatial distribution, but ultimately the pollution sources' spatial distribution. Analyzed from spatial characteristics, each pollution-discharging firm can be regarded as a pollution source, forming a point-to-surroundings dispersion pattern, while it is a linear spatial pattern if analyzed from whole river basin [9].

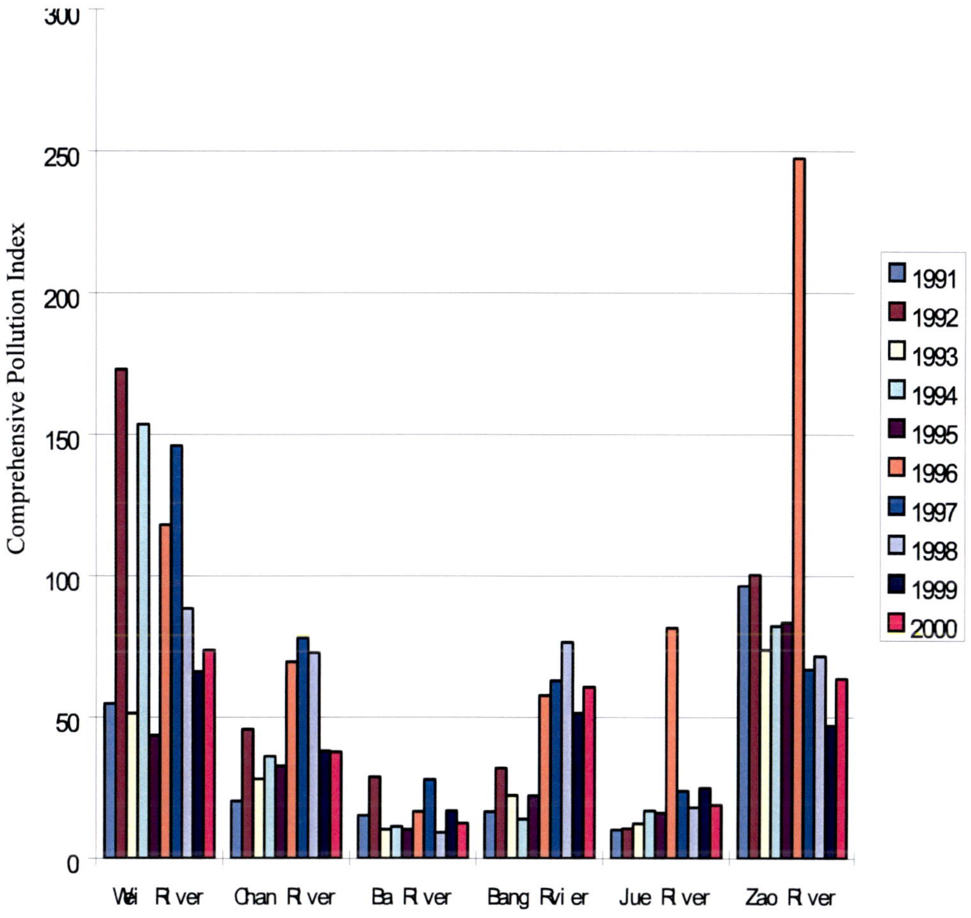

Figure 2. Comprehensive Pollution Index Change of Main Rivers in Xi'an.

2.2.2. Spatial Relationship of Rivers and Pollution-Discharging Firms in Xi'an City

From 1990's, sources of waste water pollution in Xi'an City trended to decrease in scale and increase in its amount, and discharged amount of waste water from several biggest pollution sources decreased annually after years of treatment, therefore many mid-and-small size pollution sources became the main pollution source. Papermaking industry and chemical material making industry has been the two main sources of waste water in Xi'an City and never changed. The waste water discharge amount from main pollution source is higher than 60 percent in proportion to the pollution load of the whole city. Seeking the coupling relationship between the spatial distribution of pollution sources and the pollution degree in the rivers nearby, we can easily find that they are temporally consistent and spatially interrelated (table 1).

Table 1. Properties and spatial characters of key industrial waste water pollution source in Xi'an

Year	Number of pollution discharging firms	Types of firms	Amount of waste water discharge (10^4tons)	Load-sharing coefficient	Main rivers nearby
1991	8	Chemical materials and products firms (3), Paper-making firms (2)	1910.33	72.1	Zao River, Wei River
1992	10	Chemical materials and products firms (3), Paper-making firms (3)	1819.13	61.02	Wei River, Zao River, Chan River
1993	10	Chemical materials and products firms (4), Paper-making firms (3)	2004.56	65.66	Zao River, Wei River, Chan River
1994	11	Chemical materials and products firms (4), Paper-making firms (3)	2207.94	55.81	Wei River, Zao River
1995	11	Chemical materials and products firms (4), Paper-making firms (3)	2432.78	62.55	Zao River, Wei River, Chan River
1996	23	Machinery and electronic firms (6), Chemical materials and products firms (6), Paper-making firms (3), Medical firms (3)	2538.74	70.28	Zao River, Wei River, Jue River, Feng River, Chan River
1997	24	Chemical materials and products firms (4), Paper-making firms (12), Medical firms (2)	2148.92	75.19	Wei River, Chan River, Zao River, Feng River
1998	38	Machinery and electronic firms (8), Chemical materials and products firms (5), Paper-making firms (16), Textile firms (4)	3376.88	75.47	Wei River, Feng River, Chan River, Zao River
1999	36	Paper-making firms (22), Textile firms (2), Chemical materials and products firms (2), Machinery and electronic firms (3), Electric and gas firms (2)	3923.93	75.54	Wei River, Feng River, Zao River, Chan River
2000	16	Paper-making firms (11), Chemical materials and products firms (3)	4480.93	75.67	Wei River, Feng River, Zao River, Chan River

Notice: parts of the data are from Environment Quality Report in Xi'an (1991~1995) and (1996~2000).

In a word, analyzed from the spatial morphology of the interaction between urbanization and environment, river pollution resembles to linear type, that is, linear spreading pattern, which caused the decay effect and encroachment effect.

2.3. Spatial Effect of Ground Subsidence

Geological disaster, among all kinds of natural disasters, is crucial to Xi'an City, where ground subsidence and ground fissure becomes the obstacle of urbanization there.

2.3.1. Characters of Ground Water and Its Exploitation in Xi'an

It is the imbalance of ground water spatial distribution and over-exploitation of ground water in some area that incurred the ground subsidence. In Guanzhong region, ground water and exploitable ground water are relatively abundant because of enough water-feeding amount. Ground water distributes mainly in three counties in the south of Xi'an City, 45.5 percent in water resource amount, while it is 35.4 percent in urban area. As to the exploitation amount, the former is 35 percent and the latter is 49.6 percent. The average exploitation degree of the former is 78 percent but it is 135.7 percent in urban area.

2.3.2. Characters of Ground Subsidence in Xi'an City

Ground subsidence takes place mainly in urban area and close suburb. We illustrate it according to its subsidence amount and subsidence velocity. At the end of 1995, the area whose subsidence amount is more than 200 millimeters is 145.5 square kilometers, ranging from Yuhuazhai Village in the west to textile industrial region in the east, from Sanyao Village in the south to Xinjiamiao Village in the north. In general, total subsidence amount is greater in southeast suburb than in northwest suburb. There are seven subsidence slots in this area, whose subsidence center lies separately in Xinjiamiao Village, Xi'an Jiaotong University, Hujiamiao Village, Shapo Village, Dayanta Cross, Dongbali Village and Northwestern Polytechnic University. Total subsidence amount in most of the suburb, except in northwest suburb, is more than 600 millimeters, and the subsidence area exceeding 1000 millimeters is 41 square kilometers. More seriously, in some subsidence centers, such as Dongbali Village, Shapo Village, Dayanta Cross and Hujiamiao Village, the subsidence amount is more than 2000 millimeters, and Dongbali Village center is the most serious one, 2322 millimeters.

As to the spatial distribution of subsidence velocity, the area where subsidence velocity exceeds 100 mm/a, about 8.5 square kilometers, distributes in Dongbali Village, Dayanta Cross, Shapo Village and Hujiamiao Village, inosculating to subsidence centers. The area where subsidence velocity is between 50 and 100 mm/a, about 42.5 square kilometers, mainly distributes in north suburb, south suburb and urban area, but it is lower in north suburb, west suburb and textile industrial region, 50 mm/a (table 2).

Table 2. Basic data of the main subsidence area in Xi'an

<table>
<tr><th>Area</th><th>Population (10^4)</th><th>Subsidence velocity and subsidence amount</th><th>Main firms and enterprises</th></tr>
<tr><td>Xinjiamiao Street</td><td>3.32</td><td rowspan="3">Gt. 100mm/a
Gt. 2000mm</td><td>Colliery Machine Firms, Heavy Machine Firms, Heavy Machinery Research Institute, etc.</td></tr>
<tr><td>Xi'an Jiaotong University (Taiyilu Street)</td><td>12.37</td><td>Xi'an Jiaotong University, Xingqing Park, Substation One of the Railway Bureau, etc.</td></tr>
<tr><td>Dayanta Cross (Dayanta Street)</td><td>8.50</td><td>Dayanta Park, Chang'an Lotus Garden, Study College of the CCP, etc.</td></tr>
<tr><td>Bali Village (Changyanpu Street)</td><td>14.30</td><td></td><td>Shaanxi Normal University, Xi'an Foreign Language college, Northwest University of Political and Law, Xi'an Post and Telecom College, College of Xi'an Finance and Economy</td></tr>
<tr><td>Northwestern Polytechnic University (Zhangjia Village)</td><td>13.74</td><td></td><td>Northwestern Polytechnic University, Northwestern University, Municipal Road Bureau, Provincial Hospital</td></tr>
<tr><td>South suburb</td><td>121.34</td><td rowspan="3">50 ~ 100mm/a
Gt. 1000mm</td><td>East High-tech Industrial Zone, West High-tech Industrial Zone, Culture and Education Zone, Qujiang Holiday Park</td></tr>
<tr><td>East suburb</td><td>55.65</td><td>Military Industrial Region, Heavy Industry Region, Whole Sale and Trade Region</td></tr>
<tr><td>Urban area</td><td>52.05</td><td>Residential Region, Shopping Center, Tourism region, Working Region</td></tr>
<tr><td>North suburb</td><td>42.09</td><td rowspan="3">Lt.50mm/a
Gt. 600mm</td><td>Economic and Technical Development Zone, Relic Preserving zone, Material Storage zone</td></tr>
<tr><td>West suburb</td><td>26.58</td><td>Electronic Industrial Region, Comprehensive Industrial Region</td></tr>
<tr><td>Textile industrial region</td><td>26.13</td><td>Textile Industrial Region</td></tr>
</table>

2.3.3. Reasons of Ground Subsidence

There are two reasons to cause the ground subsidence in Xi'an City. One is over-exploiting confined water, and the other is regional tectonic activities [10].

Over-exploiting confined water can result in ground water stage descent, which will subsequently incur to condensation in strata. Before the mid of 1990s, ground water is mainly used in Xi'an City and therefore there are many firms supplying water for themselves except municipal-owned water-supplying establishment. All subsidence centers mentioned above are the area with high population and intensive firms or enterprises, where the regional ground subsidence scale and velocity are homologous to local population, the scale and number of firms and enterprises (table 2). Spatial location of subsidence centers, a cumulative result for 40 years, is temporally inheritable (table 3).

Table 3. Comparison between Water Stage Descent and Ground Subsidence in Xi'an [11]

Type of change	Stage I	Stage II	Stage III	Stage IV
Water Stage Descent	In the water stage slowly-descending phase before 1970, descending velocity of confined ground water is 0.5 ~ 0.8m/ a, and no depression zone in urban area.	In the water stage acceleration-descending phase from 1971 to 1975, descending velocity of confined ground water stage is 5 m/ a, and depression zone expanded annually after its formation in urban area.	In the water stage fast-descending phase from 1976 to 1983, descending velocity of confined ground water is 4 ~ 7 m/ a, together with the deepening of intrinsic depression zone and formation of new depression zone	From 1984, in water stage slight-descending phase in 1980s and stable phase in 1990s, descending velocity of confined water stage is 3 ~ 6m/a, and the depression zone is also in stable phase.
Ground Subsidence	Before 1971, ground subside at the speed of 2 mm/a, and there are two subsidence centers in urban area	From 1972 to 1977, ground subsidence accelerated in an average velocity of 23.5mm/a, and there are five subsidence centers	From 1978 to 1984,the average velocity is about 59.1 mm/a, and 7 subsidence centers had been formed.	From 1984 to now, subsidence velocity decelerated except in the Electronic Industrial Region.

In brief, ground subsidence resembles to multi-nuclei and sector morphology, and expands in circular-layer pattern, which caused the tree-ring effect, decay effect and encroachment effect.

3. Approaches to the Coordination of Interaction between Urbanization and Environment

3.1. Coordination Hierarchies: The Coordination between Macroscopic and Microscopic Interaction

Coordination can be carried out in macroscopic, mesoscopic and microscopic hierarchies. Macroscopic coordination refers to allocation and strategies, laying emphases on urban collective allocation and its function zone, and abatement of pollutants to realized rational emission. Sound allocation can alleviate the occupation of natural resources to release the internal pressure in region. Simultaneously, it is also important to choose an urban development direction and to make a sound allocation of industries and infrastructure. Macroscopic coordination emphasize on long-term goals and strategies, relationship between city and its surrounding region, velocity and direction of urbanization, fundamental strategies on environment optimization, selection of leading industries and their development, etc. Mesoscopic coordination refers to planning and countermeasures.

Table 4. Hierarchies of the Coordination between Urbanization and Environment

	Macroscopic hierarchy (Regional or municipal scale)	Mesoscopic hierarchy (Scale in function zone or industry)	Microscopic hierarchy (Scale in community or firm)
Pollution type	Ground water and surface water pollution, Solid waste, Air pollution, Soil pollution	Sewage or waste water, waste gases, and soil pollution	Sewage or waste water
Pressure on resource	Ground water stage descent, vegetation demolishment, wetlands shrinkage, decrease of farmland, water shortage for its quality	Ground water stage descent, decrease of farmland, water shortage for its quality	Green land, Water sources
Problems of disasters	Water and soil erosion, Ground subsidence, Ground fissure, Flood and water logging, Debris flow, Epidemics, etc.	Ground subsidence, Ground fissure	Ground environment, and Construction groundwork
Cycling scale	The whole region, regardless of in the surface or under ground	Function zone	Community
Trends of expansion	The whole region	Function zone or the whole city	Community, Function zone

Studying on the function of system and its subsystem from many angles such as industry, space, environment, resources and population, Mesoscopic coordination gives prominence to quantitative analysis and recent and long-term regional planning, concretely analyze the morphology of allocation, coordination pattern between industries and environment, countermeasures of environment optimization, etc. Microscopic coordination refers to design and measure, dealing with coordination problems of a city or a certain environmental element or resource, such as pollutant emission and treatment, industry development and urban environment, patterns of urbanization, and the relationship between pollution source and industrial development.

3.2. Basic Ways of Coordination and Approaches of the Development

3.2.1. Global View as a Substitute of Partial View in Development Mechanism

Global view can be interpreted into three aspects. Firstly, we should take city as an integral system when discussing urban development problems of economy, society, culture, environment, or resource, etc. Secondly, city and environment are the two intimately related subsystems, just as an indivisible entity. The declination of either of them will surely be an obstacle to the development of the other, resulting in the entity's declination eventually. Thirdly, it is a best choice to organically combine urbanization patterns of many mechanism types to promote regional or network urbanization, because urban development is not merely the expansion in scale.

3.2.2. Diversity as a Substitute of Singularity in Development Structure

Diversity primarily refers to the three urbanization types, population concentration to city, suburbanization and population concentration to town. Diversity secondly refers to the types of urbanization, including urbanization brought by industrialization, or driven by project, or brought by tertiary industry, etc. either of which should be suitable to local condition.

3.2.3. Intensive Development as a Substitute of Extensive One

It is the premise of utilizing resources intensively to each of the urbanization types in Northwest China, including intensive urban land use, urban industries with intensive water usage, sufficient employment in urbanization, maximum products in urban firms, etc. Here we lay emphasis on the quality but not the quantity, and on the coordination of urbanization and environment.

3.2.4. Eco-Economic Goal as a Substitute of Economic Goal

Substantial forms or non-substantial forms in urbanization are multi-target restricted, including resources occupation, industrial structure and industrial type, general characters of urban system, etc. all of which should be related according to eco-economic development pattern. Simultaneously, the results and ends of urbanization should be tested by a criterion, the unification of economic benefits, ecological benefits and social benefits.

References

[1] He Mingjun, 1993. General Theories on City Development. *Urban Problems*. **1**: 17-21.

[2] BATTYM, LONGLEYPA, 1987. *Urban Shapes as Fractals. Area*.**19**: 215-221.

[3] Liu Yaolin, Liu Yanfang, Liang Qin'ou. 1999. *Urban Environment Analyses*. Wuhan Mapping Technology Press, Wuhan. 55-57, 119-120, 233.

[4] Chen Youqi. 2000. Mechanism of Landuse Evolution in Urban-rural Ectone. *Journal of China Agricultural Resources and Regional Planning*. **21**(5): 34.

[5] Han Yuanfeng, Zhang Zhixun, Zhao Ruzhi. 1993. *Regional Geographical Theories and Methods*. Shaanxi Normal University Press, Xi'an. 9-13, 127-128, 203-204, 211-213.

[6] Yang Xiaobo, Wu Qingshu, etc. 2000. *Urban ecology*. Science Press, Beijing. 119, 122, 171-172.

[7] Yuan Renmao, Yang Xiaoyan, Li Shude. 2001. Water and Soil Erosion and its Type System. *Beijing University Periodical* (Natural Science Version). **37**(3): 401-406.

[8] Lin Jun. 1997.Discussion on Water and soil Erosion in Shenzhen City. *Water and Soil Conserversion in China*. **1**: 33-35.

[9] Hellawell J M. 1986. Biological Indicators of Fresh Water Pollution and Environmental Management. *Elsevier*, London. 87.

[10] Petak W. J., Atkission A. A. Natural Hazard Risk Assessment and Public Policy. New York: *Springer-VerlagInc*. 1982.

[11] Yi Xuefa, Su Gang, Wang Weidong. 1997. Basic Characters and Formation Mechanisms of Ground Fissure Belt in Xi'an. *Earthquake and Geology*. **19**(4): 289-295.

In: Focus on Urbanization Trends
Editor: Laura M. De Smet

ISBN: 978-1-60456-109-8

Chapter 8

MEASURING URBANIZATION IN LOW INCOME COUNTRIES FOR PUBLIC HEALTH INTERVENTIONS

Andrew J. Tatem[*]
Spatial Ecology and Epidemiology Group, Tinbergen Building,
Department of Zoology, University of Oxford, South Parks Road, Oxford, OX1 3PS, UK.
Malaria Public Health and Epidemiology Group,
Centre for Geographic Medicine, KEMRI/Wellcome Trust Research Laboratories,
P.O. Box 43640, 00100 Nairobi, Kenya

Abstract

Within the next 30 years, the global proportion of urban dwellers will rise from under half to two thirds of the world's population. Such a shift will entail massive public health consequences, and most of this urban transition will occur in low-income regions of the world. Urban populations face very different health risks compared to those in rural areas and, therefore, to target effective and relevant public health interventions, the need for clear, consistent data on urbanization pattern and extent is paramount. Decision makers are increasingly seeing satellite remote sensing and geographical information systems (GIS) as cost-effective solutions to monitoring urbanization at a range of spatial scales. This overview focuses on the progress made within the fields of remote sensing and GIS in terms of mapping, monitoring, and modeling urban environments for the effective targeting of limited public health resources across the developing world. Existing challenges, drawbacks, and future prospects are all discussed, along with specific examples that demonstrate how the creation of accurate urbanization maps aid the refinement of malaria transmission and outcome estimates across Africa.

[*] E-mail address: andy.tatem@zoo.ox.ac.uk. Tel/Fax: +44 (0)1865 271262; Correspondence: Dr Andrew J Tatem, Spatial Ecology and Epidemiology Group, Tinbergen Building, Department of Zoology, University of Oxford, South Parks Road, Oxford, OX1 3PS, U.K.

Introduction

Ninety percent of global urbanization is anticipated to be concentrated in low-income countries (United Nations Population Division, 2005). Figure 1(a) shows projected urbanization trends for 2005-2030 for regions of the World. Of the countries classified as ‘less developed’ by the United Nations, 56 % of the population are forecasted to be living in urban areas in 2030, compared to 43% today. Figure 1(a) shows that almost all projected World population growth is forecast to occur in the less developed, or ‘low income’ regions. The most rapid rural-urban transitions are predicted to be seen in Africa and Asia (Figure 1(b)), as they accelerate closer to the urbanization levels seen on other continents. In Africa, just 38% of the 906 million inhabitants were urban dwellers in 2005. This is projected to increase to over 50% by 2030 as virtually all of Africa’s population growth will be concentrated in urban areas (United Nations Population Division, 2005, 2006).

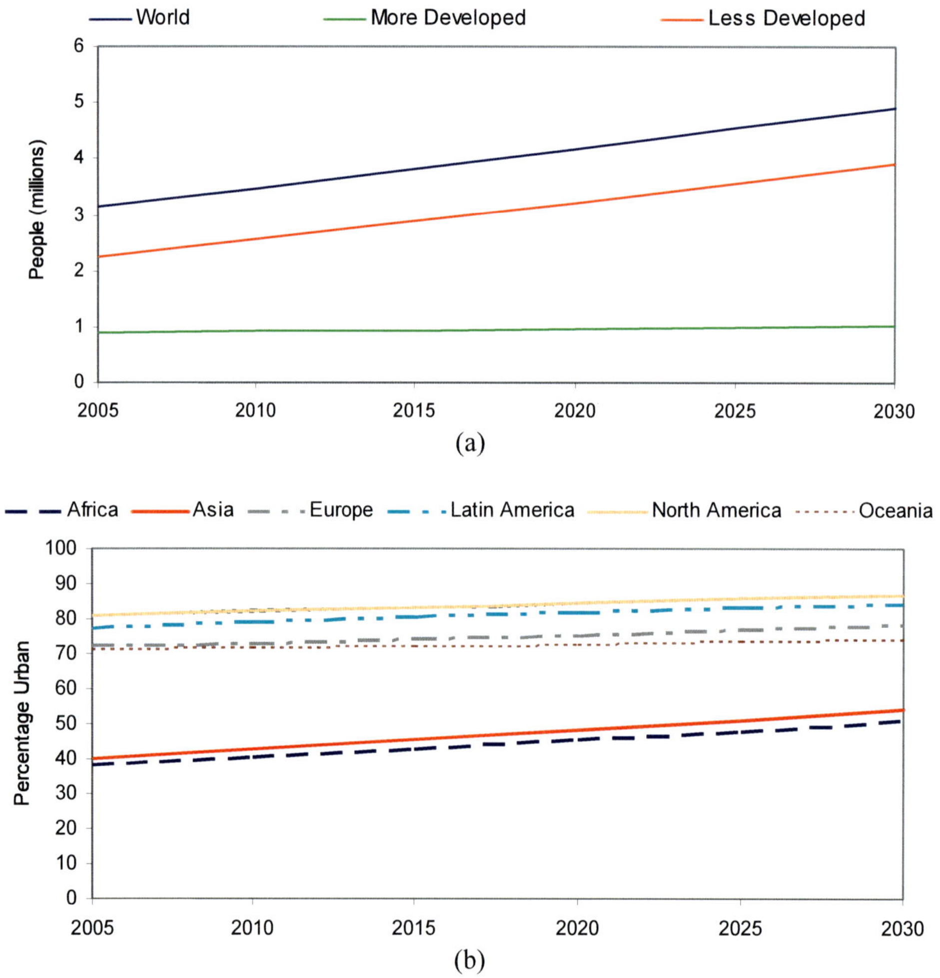

Figure 1. (a) *Global* predicted population trends 2005-2030; (b) Global predicted urbanization trends 2005-2030. (Source data: United Nations Population Division, 2005, 2006).

The profound development and epidemiological impacts of such urban-rural shifts are increasingly being realised (Campbell et al., 2007; Harpham, 1997; Harpham and Tanner, 1995; McMichael, 2000). Urban dwellers face a very different set of health risks compared to their rural counterparts (Tatem and Hay, 2004). In Sub-Saharan Africa (SSA) for example, urban residents are more at risk from directly transmitted diseases such as tuberculosis (Floyd et al., 2002) and human immunodeficiency virus (HIV) (Abebe et al., 2003; Lagarde et al., 2003) as well as certain vector-borne diseases such as dengue fever (Lines et al., 1994), but contrastingly are around 10 times less likely to receive a malaria-infected mosquito bite (Hay et al., 2005; Robert et al., 2003) and have significantly better access to health care facilities (Noor et al., 2004). Planning for the health consequences of urbanization in the countries of SSA relies on the provision of information and maps of settlement location, size and distribution.

Whilst high-income countries often have mapping resources at their disposal to define and delineate settlements for such public health planning and management, this is often not the case for low-income regions of the world. Satellite remote sensing offers a cheap and effective solution to mapping settlements and monitoring urbanization at a range of spatial scales.

Remote Sensing Data for Urban Studies

The era of satellite remote sensing began fifty years ago, on 4 October 1957, when the former Soviet Union launched Sputnik 1, the world's first artificial satellite. Sputnik 1 circled Earth once every 96 minutes and transmitted radio signals that could be received on Earth, providing the first views from space of our planet's surface and atmosphere. Since then, hundreds of Earth observing satellites have been launched, beaming back fifty years worth of imagery that have provided iconic views and unprecedented scientific insights into the workings of the planet we inhabit and our impacts on it. Satellite imagery is now familiar and accessible to all, through three-dimensional geographical information systems (GIS) such as Google Earth (http://earth.google.com/) and Microsoft Virtual Earth (http://www.microsoft.com/virtualearth/default.mspx). These allow many to view their towns and houses, as well as, those of their presidents and prime ministers. In addition to generating stunning images, the development of satellite remote sensing has facilitated substantial scientific advances over the last half century, allowing researchers to not only view and monitor the state of the Earth, but to understand and model its terrestrial, aquatic and atmospheric processes, and the impact humans have upon them.

Satellite sensors which measure the reflected radiation from the Earth's surface (Optical, passive sensors) have historically provided data for the majority of remote sensing based urban studies. The scale and high heterogeneity of reflectances across the electromagnetic spectrum compared to other land covers, means that such studies have traditionally relied on medium to high spatial resolution imagery (30m or finer). Researchers have often made use of the fact that the surface composition of urban areas has a unique reflectance signature in the visible and infrared sections of the electromagnetic spectrum to delineate and map such areas (Coiner and Levine, 1979). The application of optical sensors is often constrained by weather conditions (especially cloud), however, as well as tradeoffs between spectral, spatial and temporal resolutions, meaning increasingly the benefit of incorporating imagery from other types of satellite sensor is being seen (Cao and Jin, 2007; Tatem et al., 2004b, 2007). The majority of low income countries are in tropical regions where cloud cover prevails (New et al., 2002), therefore, imaging radars (active sensors) are finding application in aiding the monitoring and mapping of urban areas due to their independence of weather conditions or time of day/night. Such sensors hold a significant advantage over aerial photography and visible and infrared radiation based satellite sensors for urban analysis in this sense (Henderson and Xia, 1997). While passive satellite sensors measure reflectance in the visible and infrared regions that is primarily related to the molecular resonance of surface materials, the backscatter of radar waves is determined by the physical properties of surface objects such as surface roughness (Dong et al., 1997; Xia and Henderson, 1997). Radar systems can therefore add additional discriminating power making them a valuable extra tool for urban detection and analysis, and consequently public health management and planning. The detection of human settlements and detailed urban land cover and use analysis based on radar imagery is an under-developed and complex field, however, and only now, with the advent of high spatial resolution systems (e.g., Radarsat-2, Cosmo/Sky Med and TerraSAR) is it starting to receive adequate attention (Amarsaikhan et al., 2007; Dell'Acqua and Gamba, 2006; Quartulli and Tupin, 2003; Roth, 2003)

In addition to the traditional use of medium-high spatial resolution optical imagery and the more novel application of radar imagery for urban studies, the US Air Force Defense Meteorological Satellite Program (DMSP) Operational Line Scanner (OLS) has found application in large scale urban mapping (Elvidge et al., 2003; Small et al., 2005). Originally designed for night-time cloud detection, the sensor signal is intensified at night for detection of faint emission sources, resulting in city lights, fires and flares detection. Three types of 'human settlement' night-lights data have been developed (Elvidge et al., 2003): (i) stable lights, which define location and uncalibrated brightness, (ii) radiance, which is calibrated brightness, and (iii) night-time light change, which indicate temporal changes. In each case however, large numbers of night-time imagery segments are required to generate cloud-free composites. These must then be filtered to remove noise and lights from fires and other non-human settlement sources. The coarse spatial resolution and uncertainty associated with night time light size, magnitude and location (especially the 'blooming' or 'overglow' effect (Elvidge et al., 2007; Tatem et al., 2005)) means that the imagery often finds greatest utility when combined with other data sources, rather than as sole input to mapping studies, e.g., (Balk et al., 2006; Dobson et al., 2003; Schneider et al., 2003).

Identifying and Defining Urban Extents

Across the low income regions of the world, basic information on the size and extent of urban areas is often non-existent, outdated or inaccurate (Barnsley et al., 2001; Baudot, 2001). For the planning of healthcare and interventions, regularly updated information on this is important for defining populations at risk, commodity needs estimation, accessibility to facilities and intervention strategies. A wide range of image processing techniques can be applied to satellite sensor imagery to yield information on urban extents (Weber, 2001).

The identification and delineation of urban areas using satellite sensors has been studied for almost as long as the existence of Earth observing satellites. Significant early work examined the use of Landsat's 30 m spatial resolution Thematic Mapper (TM) in obtaining urban area measurements (Forster, 1983, 1985). This has since been followed up with more sophisticated approaches, ranging from simple indices (Gao and Skillcorn, 1998; Zha et al., 2003), to regional characterisations (Civco et al., 2002; Vogelmann et al., 1998), and complex data fusion approaches using Bayes modelling and Markov random fields (Yu et al., 1999). Techniques for mapping urban areas have further progressed to the extraction of statistical measures. These include deriving simple measures of urban morphology from satellite sensor imagery (Longley and Mesev, 2001; Mesev et al., 1995; Webster, 1996), through to the use of texture information (Pesaresi and Bianchin, 2001), geostatistics (Brivio and Zilioli, 2001) and a range of classifiers (Huang et al., 2007) for urban structure and pattern characterisation.

As described earlier, alternative satellite sensors are increasingly being used in urban mapping. The additional discriminating power of imaging radar over optical sensors has lead to its use in imaging urban areas (Dell'Acqua and Gamba, 2006; Hepner et al., 1998; Taket et al., 1991). DMSP OLS night-time imagery has also found use in urban area delineation, with city lights imagery used to estimate settlement area (Imhoff et al., 1997; Small et al., 2005), and combined with MODerate resolution Imaging Spectroradiometer (MODIS) imagery to improve classification accuracy of urban land cover (Schneider et al., 2003). Finally, the advent of numerous new high spatial resolution satellite sensors has prompted a boom in research aimed at quantifying the complex makeup of urban land cover, often overlooking the need and utility of such sensors to public health in simple urban extent delineation. Similar work includes the identification of informal settlements (Dare and Fraser, 2001; Weber and Puissant, 2003), and the monitoring of refugee camps (Bjorgo, 2000; Giada et al., 2003), both of which explore potentially valuable approaches for the mapping of areas with notorious public health issues (Dyke, 1991; Rowland and Nosten, 2001).

Classifying Urban Land Covers and Uses

The range of land covers within urban areas and the range of scales across which they vary, makes the classification of types of urban land cover and land use a tricky proposition (Cadenasso et al., 2007). While classified satellite imagery at a range of scales has long been used with great success in rural areas to analyse, for instance, problems of agriculture, forestry and land use, urban areas add sufficient environmental complexity to create problems for remotely sensed image analysis (Clapham, 2003). For the majority of land use surveys carried out by city authorities, maps, aerial photographs and geographical information system (GIS) layers are used, but increasingly remotely sensed imagery is being adopted. Difficulties with land use derivation from remotely sensed imagery, however, still restrict its potential as sole input to land use surveys.

Urban areas are very heterogeneous in relation to surrounding rural areas, resulting in virtually every pixel of even high spatial resolution imagery being composed of a mixture of a range of different surfaces. Moreover, identifying accurately which pixel matches which area on the ground can represent a difficult problem, leading to geo-registration errors. Inaccuracies produced when converting levels of reflectance to land cover classes, are further compounded when inferring land use from land cover maps. Further, the actual land use nomenclature used can add to these inaccuracies (Donnay and Unwin, 2001). Error and uncertainty produced from aspects such as these represent a major drawback to the operational application of remote sensing in urban land cover/use mapping (Mesev, 2003).

The difficulties described so far in inferring urban land cover and use from remotely sensed imagery can be overcome, however, and the field of urban land cover and land use classification has a long history of development. Algorithms designed to classify urban land cover and land use have developed rapidly in tandem with an ever expanding range of satellite sensors (Moller-Jensen, 1997; Neeck et al., 2005). Traditional per-pixel classifiers based on maximum likelihood theory remain popular (Baraldi and Parmiggiani, 2000; Ben-Dor and Saaroni, 2001; Mesev et al., 2001), but such approaches require training data that is normally distributed, something which cannot always be assumed in such heterogeneous environments. Techniques which make no such assumptions often prove to be more accurate, e.g., neural networks (Gamba and Houshmand, 2001; Paola and Schowengerdt, 1995), support vector machines (Brown et al., 2000) or wavelet-based approaches (Myint, 2006). Taking things further, often image pixel size is too coarse for accurate urban land cover representation, so recent research has advanced sub-pixel classification using mixture models (Phinn et al., 2002), neural networks (Gamba and Dell'Acqua, 2003; Zhang and Foody, 1998) and even multi-stage approaches to produce maps at finer resolutions than the input imagery (Nguyen et al., 2006; Tatem et al., 2001a, b, 2003).

Urban land cover is spatially complex and, dependent upon the type of imagery used, often spectrally mixed within pixels. Classification accuracy may therefore only improve in some cases if the spatial properties of each class can be incorporated into the classification (Herold et al., 2003b; Jensen, 1979). This can be undertaken via the inclusion of image texture information, typically extracted via either region-based or window-based approaches (Haralick et al., 1973) More complex spatial, contextual and structural pattern recognition techniques are also finding application in urban land use classification (Barnsley et al., 2001). In addition to these, attempts at incorporating ancillary georeferenced datasets are also proving successful (Harris and Ventura, 1995; Kam, 1995). With GIS increasingly finding a place within public health monitoring and planning, approaches which enable extra data sources to be incorporated to improve and expand the information content of satellite-derived urban zonations should be welcomed, e.g., (Harris and Ventura, 1995; Stefanov et al., 2001).

Measuring Urbanization

The rapid industrialization and urbanization in many low income countries is potentially leading to long-term detrimental public health effects (Campbell et al., 2007; Harpham and Tanner, 1995; Harris and Ventura, 1995; McMichael, 2000; McMichael and Beaglehole, 2000). This has resulted in a need for evaluation of the magnitude, pattern and type of land cover and land use changes, and for prediction of future urban development (Weng, 2002). This need could be met by satellite remote sensing, however, as discussed above, the spatial and spectral heterogeneity of urban surfaces, and rapid change in radiance levels occurring over short time periods means that the analysis and interpretation of urban change from satellite imagery is challenging. Integrating remotely sensed imagery with other sources, such as socio-economic data presents a further challenge through differences in the nature of data collection objectives, units of analysis, sources of error and spatial resolution (Jensen and Cowen, 1999).

The research field of urban change detection has a relatively long history, and simplistic early work typically fell into one of two categories: (a) post-classification comparison and (b) temporal sequence composition. The former technique often involves a basic per-pixel land cover classification of two temporally different images of identical location, which are then differenced to identify the magnitude of changes and the alteration of land cover (Batty and Howes, 2001; Braimoh and Onishi, 2007; Kressler and Steinnocher, 1996; Tang et al., 2006). The latter approach simply involves building up a geo-registered composite of temporally differing images from the same sensor. Visual comparison, differencing (Ji et al., 2001; Kwarteng and Chavez, 1998) or applications of a range of change detection algorithms (Ridd and Liu, 1998; Yang et al., 2003; Zhang, 2001) are then applied to quantify changes. Such techniques are ideal for identifying largescale changes over time, but their simplicity means that often small changes are overlooked, and no real understanding of the process involved and potential future changes can be gleamed.

Recent years have seen more sophisticated approaches to using satellite remote sensing for examining the spatial and temporal patterns of land cover and land use change (Herold et al., 2003a). These have been aimed predominantly at quantifying the magnitude, location and type of urban change, but some models have been developed to address how and why changes have occurred, and project such changes into the future. Two categories of such models exist: regression-based and spatial transition-based (Theobold and Hobbs, 1998). The majority of recent studies have followed the regression-based approach and simply relate land cover/use change locations to sets of spatially explicit variables (Liu and Lathrop, 2002; Qiu et al., 2003; Seto and Weiguo, 2003). Spatial transition-based models are designed to predict future urban changes based on probabilistic estimates from Monte Carlo, Markovian or other models (Clarke and Gaydos, 1998; Wu et al., 2006). Such models can take the form of cellular automaton simulations, although few studies have successfully incorporated remotely sensed imagery as a principal model driver. The direction of future research in the field looks to be towards more complex models such as these, which incorporate both new and additional types of imagery, e.g., night-time imagery (Doll, 2003), synthetic aperture radar (Gamba et al., 2006) and radar interferometry (Grey et al., 2003).

An Example: Urban Areas and Global Malaria Burden

The last attempt to map malaria risks worldwide was in the 1960s (Lysenko and Semashko, 1968). For many areas of the world this, unfortunately, still represents our best information on malaria risk. The distribution of populations exposed to the risk of *Plasmodium falciparum* malaria is also poorly understood. Without an understanding of the spatial distribution of malaria endemicity and populations exposed to the disease, it is not possible to develop credible, evidence-based estimates of disease burden. It is also not possible to set clear international priorities for malaria prevention and control.

Considerable effort is therefore required to improve basic maps of malaria endemicity. Environmental changes (climate, population growth, landcover and urbanization) have and will continue to affect the epidemiology of malaria. These influences must be incorporated into any present day estimates or temporal projections of future malaria risk to define non-intervention changes in denominator populations, otherwise objective measures of our achievements toward international development goals for malaria will be impossible (Hay and Snow, 2006).

As a general rule, cities are unhealthy for the malaria parasite. There is clear evidence that urbanization affects anopheline species in the environment — diversity, numbers, survival rates, infection rates with *P. falciparum* and the frequency with which they bite people are all affected (Hay et al., 2005). So, fewer people acquire malaria infection, become ill and/or die of its consequences in urban areas. The most common explanation is lower vector densities that result from a paucity of clean freshwater breeding sites (Lindsay et al., 1990). As has been eloquently detailed (Trape and Zoulani, 1987), however, the process of urbanization effects changes in indices of mosquito and malaria abundance not only by eliminating open spaces for breeding, but also by increasing pollution of the remaining breeding sites, thereby limiting the dispersion opportunities for adult mosquitoes. With increased human densities, malaria exposure per capita also decreases (Smith et al., 2004; Trape and Zoulani, 1987).

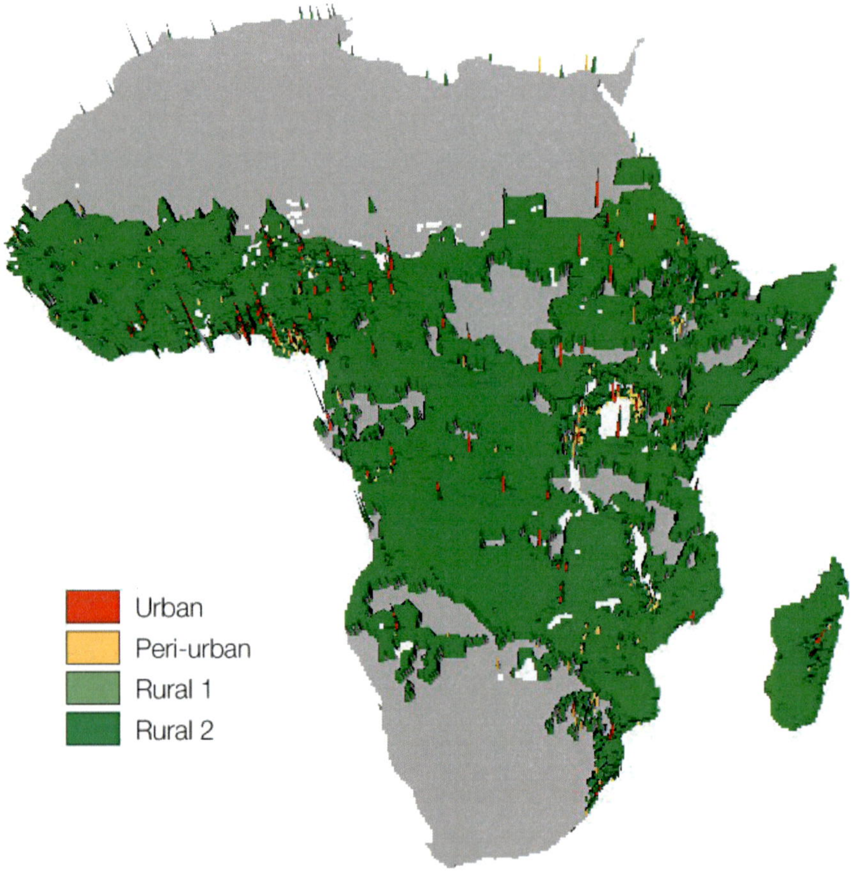

Figure 2. 3d representation of the GRUMP urban extent surface combined with gridded population density data and malaria transmission limits to show populations at risk. The grey areas show either areas of population density less than one person per km^2 or of predicted zero malaria transmission. Adapted from Hay et al. (2005).

To improve present and future maps of malaria endemicity, malaria burden estimates and consequently, estimates of intervention requirements, accurate estimates of the present extent of urban areas and future urbanization in malaria endemic countries are required. Initial steps towards achieving this have been so far based upon DMSP OLS night-time light satellite imagery for Africa (Hay et al., 2005; Keiser et al., 2004). Whilst such imagery provides a consistent largescale view, it does however suffer from problems in terms of mapping urban areas across low income regions. Firstly, the blooming effect described previously produces urban extent overestimation (Tatem et al., 2005), and while this can be controlled through image thresholding, this makes the second problem worse: the low electrification of many significant settlements across Africa means they remain undetected. Steps to alleviate such issues have been taken in the production of an urban extent mask for the Global Rural Urban Mapping Project (GRUMP: http://sedac.ciesin.org/gpw/), which combines night time lights imagery with various other layers to ensure lowly-lit settlements are mapped (Balk et al., 2006). This mask was used in combination with population densities and malaria transmission limit rules (figure 2) to build some of the first continent-wide relationships between malaria endemicity measures and urbanization, and consequently produce urban-corrected estimates of malaria risk for Africa (Hay et al., 2005). These efforts have since been extended to global burden estimates (Snow et al., 2005), transmission limits (Guerra et al., 2006) and future burdens (Hay et al., 2006), whilst the Malaria Atlas Project (MAP: http://www.map.ox.ac.uk) (Hay and Snow, 2006) aims to refine and improve upon these.

Future Prospects

The last fifty years has seen satellite-derived remotely sensed imagery become a central data source for many scientific fields, including both public health and urbanization. The future is likely to see the influence of remote sensing within these research fields grow, as satellite imagery becomes cheaper, more widely available, more accessible and of higher quality. The continuing increase in computing speeds is likely to make the processing of largescale, multiband satellite imagery easier, whilst decreasing computing costs and the rise of open-source GIS software should mean wider accessibility to the data, tools and benefits of urban mapping for public health applications.

Two divergent trends in Earth observation satellite commissioning and design are being seen at present: (i) large, multidisciplinary, multi-sensor, global-coverage satellites and (ii) small, fine spatial resolution, national satellites. The first type of satellite is exemplified by NASA's Terra and Aqua (Tatem et al., 2004a), and ESA's Envisat (Louet, 2001). Both carry numerous sensors and produce global imagery at an unprecedented spatial resolution. Onboard Terra and Aqua are MODIS sensors, and the imagery produced has been used to create some of the highest resolution global urban maps available, as well as constrain the blooming in night time lights imagery for urban mapping (Schneider et al., 2003). Future sensors should extend global urban analysis capabilities further, such as the planned Visible/Infrared Imager/Radiometer Suite (VIIRS) onboard the National Polar-orbiting Operational Environmental Satellite System (NPOESS) (Lee et al., 2006) and the proposed Nightsat (Elvidge et al., 2007). The second type of satellite is becoming increasingly common. Large, wealthy countries have had their own Earth observation satellites for many years. Now, with the costs of building and launching small Earth observation satellites decreasing and the benefits of land cover monitoring being increasingly recognised, more and more countries are launching their own satellites, including recently Brazil, India and Nigeria. Both such trends in Earth observing satellites are likely advance our abilities to measure urbanization in low income countries. The global-coverage satellites will provide consistent measurements across large areas to enable refinement of global urban map products, allowing us to, for example, improve the precision of disease burden estimates. Moreover, the smaller national satellites will provide, for the first time, time series of imagery with high temporal and spatial resolution focussed on single countries, facilitating detailed measurements of urbanization trends and reducing the limiting effects of cloud cover for example.

Advances in satellite sensor technology may also result in the breaking down of barriers which have hindered urban remote sensing in the past. Specifically, the traditional trade-offs between spectral, spatial and temporal resolutions are beginning to be eroded as global coverage sensors, such as MODIS, provide regular imagery at unprecedented spatial resolutions for such sensors and in greater numbers of wavebands than ever before. Such features allow improved urban land use classification, across greater areas, more regularly and in more detail than ever before. With computers continuing to advance in power to cope with the processing of such imagery, and new methodologies to process and present resulting analyses and maps being developed and made opensource, the future of urbanization monitoring from space looks bright.

Conclusion

Urbanization has significant and far-reaching public health impacts, particularly in low income countries. Monitoring, mapping and preparing for such changes over a range of scales requires consistent definitions and approaches for defining and delineating urban areas. This review has demonstrated that the progress made in using remote sensing for describing and modelling urban environments puts it in an excellent position to provide such information to the public health field and beyond. In particular, remote sensing has a central role to play in low-income areas, in which resources for mapping and healthcare are stretched, and the effects of urbanization will be felt most.

Acknowledgements

This work forms part of the output of the Malaria Atlas Project (MAP, http://www.map.ox.ac.uk), principally funded by the Wellcome Trust, U.K.

References

Abebe, Y., Schaap, A., Mamo, G., Negussie, A., Darimo, B., Wolday, D., and Sanders, E. J. (2003). HIV prevalence in 72 000 urban and rural male army recruits, Ethiopia *AIDS* **17**, 1835-1840.

Amarsaikhan, D., Ganzorig, M., Ache, P., and Blotevogel, H. (2007). The integrated use of optical and InSAR data for urban land-cover mapping. *International Journal of Remote Sensing* **28**, 1161-1171.

Balk, D. L., Deichmann, U., Yetman, G., Pozzi, F., Hay, S. I., and Nelson, A. (2006). Determining global population distribution: methods, applications and data. *Advances in Parasitology* **62**, 119-156.

Baraldi, A., and Parmiggiani, F. (2000). Urban area classification by multispectral SPOT images. *IEEE Transactions on Geoscience and Remote Sensing* **28**, 674-680.

Barnsley, M. J., Moller-Jensen, L., and Barr, S. L. (2001). Inferring urban land use by spatial and structural pattern recognition. *In "Remote Sensing and Urban Analysis"* (J.-P. Donnay, M. J. Barnsley and P. A. Longley, eds.), pp. 115-144. Taylor and Francis, London.

Batty, M., and Howes, D. (2001). Predicting temporal patterns in urban development from remote imagery. *In "Remote Sensing and Urban Analysis"* (J.-P. Donnay, M. J. Barnsley and P. A. Longley, eds.), pp. 185-204. Taylor and Francis, London.

Baudot, Y. (2001). Geographical analysis of the population of fast-growing cities in the third world. *In "Remote Sensing and Urban Analysis"* (J.-P. Donnay, M. J. Barnsley and P. A. Longley, eds.), pp. 225-241. Taylor and Francis, London.

Ben-Dor, E., and Saaroni, H. (2001). A spectral based recognition of the urban environment using the visible and near infrared spectral region. A case study over Tel-Aviv, Israel. *International Journal of Remote Sensing* **22**, 2193-2218.

Bjorgo, E. (2000). Using very high spatial resolution multispectral satellite sensor imagery to monitor refugee camps. *International Journal of Remote Sensing* **21**, 611-616.

Braimoh, A. K., and Onishi, T. (2007). Spatial determinants of urban land use change in Lagos, Nigeria. *Land Use Policy* **24**, 502-515.

Brivio, P. A., and Zilioli, E. (2001). Urban pattern characterization through geostatistical analysis of satellite images. *In "Remote Sensing and Urban Analysis"* (J.-P. Donnay, M. J. Barnsley and P. A. Longley, eds.), pp. 40-53. Taylor and Francis, London.

Brown, M., Lewis, H. G., and Gunn, S. R. (2000). Linear spectral mixture models and support vector machines for remote sensing. *IEEE Transactions on Geoscience and Remote Sensing* **38**, 2346-2360.

Cadenasso, M. L., Pickett, S. T. A., and Schwarz, K. (2007). Spatial heterogeneity in urban ecosystems: reconceptualizing land cover and a framework for classification. *Frontiers in Ecology and the Environment* **5**, 80-88.

Campbell, M., Cleland, J., Ezeh, A., and Prata, N. (2007). Return of the population growth factor. *Science* **315**, 1501-1502.

Cao, G., and Jin, Y. Q. (2007). A hybrid algorithm of the BP-ANN/GA for classification of urban terrain surfaces with fused data of Landsat ETM+ and ERS-2 SAR *International Journal of Remote Sensing* **28**, 293-305.

Civco, D. L., Hurd, J. D., Wilson, E. H., Arnold, C. L., and Prisloe, M. P. (2002). Quantifying and describing urbanizing landscapes in the Northeast United States. *Photogrammetric Engineering and Remote Sensing* **68**, 1083-1090.

Clapham, W. B. (2003). Continuum-based classification of remotely sensed imagery to describe urban sprawl on a watershed scale. *Remote Sensing of Environment* **86**, 322-340.

Clarke, K. C., and Gaydos, L. J. (1998). Loose-coupling a cellular automaton model and GIS: long-term urban growth prediction for San Francisco and Washington/Baltimore. *International Journal of Geographic Information Science* **12**, 699-714.

Coiner, J. C., and Levine, A. L. (1979). Applications of remote sensing to urban problems. *Urban Systems* **4**, 205-219.

Dare, P. M., and Fraser, C. S. (2001). Mapping informal settlements using high resolution imagery. *International Journal of Remote Sensing* **22**, 1399-1401.

Dell'Acqua, F., and Gamba, P. (2006). Discriminating urban environments using multiscale texture and multiple SAR images. *International Journal of Remote Sensing* **27**, 3797-3812.

Dobson, J. E., Bright, E. A., Coleman, P. R., and Bhaduri, B. L. (2003). Landscan: a global population database for estimating populations at risk. *In "Remotely Sensed Cities"* (V. Mesev, ed.), pp. 267-279. Taylor and Francis, London.

Doll, C. N. H. (2003). Estimating non-population activities from night-time satellite imagery. *In "Remotely Sensed Cities"* (V. Mesev, ed.), pp. 335-353. Taylor and Francis, London.

Dong, Y., Forster, B., and Ticehurst, C. (1997). Radar backscatter analysis for urban environments. *International Journal of Remote Sensing* **18**, 1351-1364.

Donnay, J.-P., and Unwin, D. (2001). Modelling geographical distributions in urban areas. *In "Remote Sensing and Urban Analysis"* (J.-P. Donnay, M. J. Barnsley and P. A. Longley, eds.), pp. 205-224. Taylor and Francis, London.

Dyke, T. (1991). Public health aspects of a refugee camp. *Journal of the Royal Society of Health* **111**, 101-104.

Elvidge, C., Hobson, V. R., Nelson, I. L., Safran, J. M., Tuttle, B. T., Dietz, J. B., and Baugh, K. (2003). Overview of DMSP OLS and scope of applications. *In "Remotely Sensed Cities"* (V. Mesev, ed.), pp. 281-299. Taylor and Francis, London.

Elvidge, C. D., Cinzano, P., Pettit, D. R., Arvesen, J., Sutton, P., Small, C., Nemani, R., Longcore, T., Rich, C., Safran, J. M., Weeks, J., and Ebener, S. (2007). The Nightsat mission concept. *International Journal of Remote Sensing* **28**, 2645-2670.

Floyd, K., Blanc, L., Raviglione, M., and Lee, J. W. (2002). Resources required for global tuberculosis control. *Science* **295**, 2040-2041.

Forster, B. (1983). Some urban measurements from Landsat data. *Photogrammetric Engineering and Remote Sensing* **14**, 1693-1707.

Forster, B. (1985). An examination of some problems and solutions in monitoring urban areas from satellite platforms. *International Journal of Remote Sensing* **6**, 139-151.

Gamba, P., and Dell'Acqua, F. (2003). Increased accuracy multiband urban classification using a neuro-fuzzy classifier. *International Journal of Remote Sensing* **24**.

Gamba, P., Dell'Acqua, F., and Lisini, G. (2006). Change detection of multitemporal SAR data in urban areas combining feature-based and pixel-based techniques *IEEE Transactions on Geoscience and Remote Sensing* **44**, 2820-2827.

Gamba, P., and Houshmand, B. (2001). An efficient neural classification of SAR and optical urban images. *International Journal of Remote Sensing* **22**, 1535-1553.

Gao, J., and Skillcorn, D. (1998). Capability of SPOT XS data in producing detailed land cover maps at the urban-rural periphery. *International Journal of Remote Sensing* **19**, 2877-2891.

Giada, S., De Groeve, T., Ehrlich, D., and Soille, P. (2003). Can satellite images provide useful information on refugee camps? *International Journal of Remote Sensing* **24**, 4249-4250.

Grey, W., Luckman, A., and Holland, D. (2003). Mapping urban change in the UK using satellite radar interferometry. *Remote Sensing of Environment* **87**, 16-22.

Guerra, C. A., Snow, R. W., and Hay, S. I. (2006). Mapping the global extent of malaria in 2005. *Trends in Parasitology* **22**, 353-358.

Haralick, R. M., Shanmugam, K., and Dinstein, I. (1973). Texture features for image classification. *IEEE Transactions on Systems, Man, and Cybernetics* **3**, 610-621.

Harpham, T. (1997). Urbanisation and health in transition. *Lancet* **349**, 11-13.

Harpham, T., and Tanner, M., eds. (1995). *"Urbanisation health in developing countries: progress and prospects,"* pp. 1-256. Earthscan, London.

Harris, P. M., and Ventura, S. J. (1995). The integration of geographic data with remotely sensed imagery to improve classification in an urban area. *Photogrammetric Engineering and Remote Sensing* **61**, 993-998.

Hay, S. I., Guerra, C. A., Tatem, A. J., Atkinson, P. M., and Snow, R. W. (2005). Urbanization, malaria transmission and disease burden in Africa. *Nature Reviews Microbiology* **3**, 81-90.

Hay, S. I., and Snow, R. W. (2006). The Malaria Atlas Project: devloping global maps of malaria risk. *PLoS Medicine* **3**, e473-.

Hay, S. I., Tatem, A. J., Guerra, C. A., and Snow, R. W. (2006). "*Population at malaria risk in Africa: 2005, 2015 and 2030.*" UK Ofiice of Science and Innovation, London.

Henderson, F. M., and Xia, Z.-G. (1997). SAR applications in human settlement detection, population estimation and urban land use pattern analysis: a status report. *IEEE Transactions on Geoscience and Remote Sensing* **35**, 79-85.

Hepner, G. F., Houshmand, B., Kulikov, I., and Bryant, N. (1998). Investigation of the integration of AVIRIS and IFSAR for urban analysis. *Photogrammetric Engineering and Remote Sensing* **64**, 813-820.

Herold, M., Goldstein, N. C., and Clarke, K. C. (2003a). The spatiotemporal form of urban growth: measurement analysis and modeling. *Remote Sensing of Environment* **86**, 286-302.

Herold, M., Liu, X., and Clarke, K. C. (2003b). Spatial metrics and image texture for mapping urban land use. *Photogrammetric Engineering and Remote Sensing* **69**, 991-1002.

Huang, X., Zhang, L. P., and P.X., L. (2007). Classification and extraction of spatial features in urban areas using high-resolution multispectral imagery *IEEE Geoscience and Remote Sensing Letters* **4**, 260-264.

Imhoff, M. L., Lawrence, W. T., Stutzer, D. C., and Elvidge, C. D. (1997). A technique for using composite DMSP.OLS "city lights" satellite data to map urban area. *Remote Sensing of Environment* **61**, 361-370.

Jensen, J. R. (1979). Spectral and textural features to classify elusive land cover at the urban fringe. *Professional Geographer* **31**, 400-409.

Jensen, J. R., and Cowen, D. C. (1999). Remote sensing of urban suburban infrastructure and socio- economic attributes. *Photogrammetric Engineering and Remote Sensing* **65**, 611-622.

Ji, C. Y., Liu, Q. H., Sun, D. F., Wang, S., Lin, P., and Li, X. W. (2001). Monitoring urban expansion with remote sensing in China. *International Journal of Remote Sensing* **22**, 1441-1455.

Kam, T. (1995). Integrating GIS and remote sensing techniques for urban land-cover and land-use analysis. *Geocarto International* **10**, 39-48.

Keiser, J., Utzinger, J., Caldas de Castro, M., Smith, T. A., Tanner, M., and Singer, B. H. (2004). Urbanization in sub-Saharan Africa and implication for malaria control. *American Journal of Tropical Medicine and Hygiene* **71**, 118-127.

Kressler, F., and Steinnocher, K. (1996). Change detection in urban areas using satellite images and spectral mixture analysis. *International Archives of Photogrammetry and Remote Sensing* **31**, 379-383.

Kwarteng, A. Y., and Chavez, P. S. (1998). Change detection study of Kuwait City and environs using multi-temporal Landsat Thematic Mapper data. *International Journal of Remote Sensing* **19**, 1651-1662.

Lagarde, E., van der Loeff, M. S., Enel, C., Holmgren, B., Dray-Spira, R., Pison, G., Piau, J. P., Delaunay, V., M'Boup, S., Ndoye, I., Coeuret-Pellicer, M., Whittle, H., and Aaby, P. (2003). Mobility and the spread of human immunodeficiency virus into rural areas of West Africa *International Journal of Epidemiology* **32**, 744-752.

Lee, T. E., Miller, S. D., Turk, F. J., Schueler, C., Julian, R., Deyo, S., Dills, P., and Wang, S. (2006). The NPOESS VIIRS day/night visible sensor *Bulletin of the American Meteorological Society* **87**, 191.

Lindsay, S. W., Campbell, H., Adiamah, J. H., Greenwood, A. M., Bangali, J. E., and Greenwood, B. M. (1990). Malaria in a peri-urban area of The Gambia. *Annals of Tropical Medicine and Parasitology* **84**, 553-562.

Lines, J., Harpham, T., Leake, C., and Schofield, C. (1994). Trends, priorities and policy directions in the control of vector-borne diseases in urban environments. *Health Policy and Planning* **9**, 113-129.

Liu, X., and Lathrop, R. G. (2002). Urban change detection based on an artificial neural network. *International Journal of Remote Sensing* **23**, 2513-2518.

Longley, P. A., and Mesev, V. (2001). Measuring urban morphology using remotely-sensed imagery. In *"Remote Sensing and Urban Analysis"* (J.-P. Donnay, M. J. Barnsley and P. A. Longley, eds.), pp. 163-183. Taylor and Francis, London.

Louet, J. (2001). The Envisat mission and system. *Esa Bulletin-European Space Agency* **106**, 11-25.

Lysenko, A. J., and Semashko, I. N. (1968). Geography of malaria. A medico-geographic profile of an ancient disease [in Russian]. *In* "Itogi Nauki: Medicinskaja Geografija" (A. W. Lebedew, ed.), pp. 25-146. Academy of Sciences, USSR, Moscow.

McMichael, A. J. (2000). The urban environment and health in a world of increasing globalization: issues for developing countries. *Bulletin of the World Health Organization* **78**, 1117-1126.

McMichael, A. J., and Beaglehole, R. (2000). The changing global context of public health. *Lancet* **356**, 495-499.

Mesev, V. (2003). Urban land use uncertainty. In *"Remotely Sensed Cities"* (V. Mesev, ed.), pp. 207-222. Taylor and Francis, London.

Mesev, V., Gorte, B., and Longley, P. A. (2001). Modified maximum-likelihood classification algorithms and their application to urban remote sensing. In *"Remote Sensing and Urban Analysis"* (J.-P. Donnay, M. J. Barnsley and P. A. Longley, eds.), pp. 72-88. Taylor and Francis, London.

Mesev, V., Longley, P. S., Batty, M., and Xie, Y. (1995). Morphology from imagery: detecting and measuring the density of urban land use. *Environment and Planning A* **27**, 759-780.

Moller-Jensen, L. (1997). Classification of urban land cover based on expert systems, object models and texture. *Computers, Environment and Urban Systems* **21**, 291-302.

Myint, S. W. (2006). A new framework for effective urban land use and land cover classification: A wavelet approach. *GIscience and Remote Sensing* **43**, 155-178.

Neeck, S. P., Magner, T. J., and Paules, G. E. (2005). NASA's small satellite missions for earth observation. *Acta Astronautica* **56**, 187-192.

New, M., Lister, D., Hulme, M., and Makin, I. (2002). A high-resolution data set of surface climate over global land areas. *Climate Research* **21**, 1-25.

Nguyen, M. Q., Atkinson, P. M., and Lewis, H. G. (2006). Superresolution mapping using a hopfield neural network with fused images. *IEEE Transactions on Geoscience and Remote Sensing* **44**, 736-749.

Noor, A. M., Gikandi, P. W., Hay, S. I., Muga, R. O., and Snow, R. W. (2004). Creating spatially defined databases for equitable health service planning in low-income countries: the example of Kenya. *Acta Tropica* **91**, 239-251.

Paola, J. D., and Schowengerdt, R. A. (1995). A detailed comparison of backpropagation neural network and maximum-likelihood classifiers for urban land use classification. *IEEE Transactions on Geoscience and Remote Sensing* **33**, 981-996.

Pesaresi, M., and Bianchin, A. (2001). Recognizing settlement structure using mathematical morphology and image texture. *In* "Remote Sensing and Urban Analysis" (J.-P. Donnay, M. J. Barnsley and P. A. Longley, eds.), pp. 56-67. Taylor and Francis, London.

Phinn, S., Stanford, M., Scarth, P., Murray, A. T., and Shyy, P. T. (2002). Monitoring the composition of urban environments based on the vegetation-impervious surface-soil (VIS) model by subpixel analysis techniques. *International Journal of Remote Sensing* **23**, 4131-4153.

Qiu, F., Woller, K. L., and Briggs, R. (2003). Modeling urban population growth from remotely sensed imagery and TIGER GIS road data. *Photogrammetric Engineering and Remote Sensing* **69**, 1031-1042.

Quartulli, M., and Tupin, F. (2003). Information extraction from high resolution SAR data for urban scene understanding. In *"Urban 2003"*, pp. 115-119. IEEE, Berlin.

Ridd, M. K., and Liu, J. (1998). A comparison of four algorithms for change detection in an urban environment. *Remote Sensing of Environment* **63**, 95-100.

Robert, V., MacIntyre, K., Keating, J., Trape, J. F., Duchemin, J. B., Warren, M., and Beier, J. C. (2003). Malaria transmission in urban sub-Saharan Africa. *American Journal of Tropical Medicine and Hygiene* **68**, 169-176.

Roth, A. (2003). TerraSAR-X: A new perspective for scientific use of high resolution spaceborne SAR data. In *"Urban 2003"*, pp. 4-7. IEEE, Berlin.

Rowland, M., and Nosten, F. (2001). Malaria epidemiology and control in refugee camps and complex emergencies. *Annals of Tropical Medicine and Parasitology* **95**, 741-754.

Schneider, A., Friedl, M. A., and Woodcock, C. E. (2003). Mapping Urban Areas by Fusing Multiple Sources of Coarse Resolution Remotely Sensed Data. *In* "International Geoscience and Remote Sensing Symposium". *IEEE, Toulouse, France.*

Seto, K. C., and Weiguo, L. (2003). Comparing ARTMAP neural network with the maximum-likelihood classifier for detecting urban change. *Photogrammetric Engineering and Remote Sensing* **69**, 981-990.

Small, C., Pozzi, F., and Elvidge, C. (2005). Spatial analysis of global urban extent from DMSP-OLS night lights *Remote Sensing of Environment* **96**, 277-291.

Smith, D. L., Dushoff, J., and McKenzie, F. E. (2004). The risk of a mosquito-borne infection in a heterogenous environment. *PLoS Biology* **2**, e368.

Snow, R. W., Guerra, C. A., Noor, A. M., Myint, H. Y., and Hay, S. I. (2005). The global distribution of clinical episodes of *Plasmodium falciparum* malaria. *Nature* **434**, 214-217.

Stefanov, W. L., Ramsey, M. S., and Christensen, P. R. (2001). Monitoring urban land cover change: An expert system approach to land cover classification of semiarid to arid urban centers. *Remote Sensing of Environment* **77**, 173-185.

Taket, N. D., Howarth, S. M., and Burge, R. E. (1991). A model for the imaging of urban areas by synthetic aperture radar. *IEEE Transactions on Geoscience and Remote Sensing* **29**, 432-443.

Tang, J. M., Wang, L., and Yao, Z. J. (2006). Analyzing urban sprawl spatial fragmentation using multi-temporal satellite images *GIscience and Remote Sensing* **43**, 218-232.

Tatem, A. J., Goetz, S. J., and Hay, S. I. (2004a). Terra and Aqua: new data for epidemiology and public health. *International journal of Applied Earth Observation and Geoinformation* **6,** 33-46.

Tatem, A. J., and Hay, S. I. (2004). Measuring urbanization pattern and extent for malaria research: a review of remote sensing approaches. *Journal of Urban Health* **81**, 363-376.

Tatem, A. J., Lewis, H. G., Atkinson, P. M., and Nixon, M. S. (2001a). Super-resolution mapping of urban scenes from IKONOS imagery using a Hopfield neural network. In *"International Geoscience and Remote Sensing Symposium"* (IEEE, ed.). IEEE, Sydney, Australia.

Tatem, A. J., Lewis, H. G., Atkinson, P. M., and Nixon, M. S. (2001b). Super-resolution target identification from remotely sensed images using a Hopfield neural network. *IEEE Transactions on Geoscience and Remote Sensing* **39**, 781-796.

Tatem, A. J., Lewis, H. G., Atkinson, P. M., and Nixon, M. S. (2003). Increasing the spatial resolution of Landsat TM imagery for land cover mapping in agricultural areas. *International Journal of Geographic Information Science* **17**, 647-672.

Tatem, A. J., Noor, A. M., and Hay, S. I. (2004b). Defining approaches to settlement mapping for public health management in Kenya using medium spatial resolution satellite imagery. *Remote Sensing of Environment* **93**, 42-52.

Tatem, A. J., Noor, A. M., and Hay, S. I. (2005). Assessing the accuracy of satellite derived global and national urban maps in Kenya. *Remote Sensing of Environment* **96**, 87-97.

Tatem, A. J., Noor, A. M., von Hagen, C., di Gregorio, A., and Hay, S.I. (2007). High resolution population maps for low income nations: combining land cover and census in East Africa. *PLoS ONE* **2**, e1298.

Theobold, D. M., and Hobbs, N. T. (1998). Forecasting rural land use change: a comparison of regression and spatial transition-based models. *Geographical and Environmental Modelling* **2,** 65-82.

Trape, J. F., and Zoulani, A. (1987). Malaria and urbanisation in Central Africa: the example of Brazzaville. Part II. Results of entomological surveys and epidemiological analaysis. *Transactions of the Royal Society of Tropical Medicine and Hygiene* **81**, 34-42.

United Nations Population Division (2005). "World urbanization prospects, 2005 revision." United Nations, New York.

United Nations Population Division (2006). "World population prospects, 2006 revision." United Nations, New York.

Vogelmann, J. E., Sohl, T., and Howard, S. M. (1998). Regional characterization of land cover using multiple sources of data. *Photogrammetric Engineering and Remote Sensing* **64**, 45-57.

Weber, C. (2001). Urban agglomeration delimitation using remote sensing. *In* "Remote Sensing and Urban Analysis" (J.-P. Donnay, M. J. Barnsley and P. A. Longley, eds.), pp. 145-159. Taylor and Francis, London.

Weber, C., and Puissant, A. (2003). Urbanization pressure and modeling of urban growth: example of the Tunis metropolitan area. *Remote Sensing of Environment* **86**, 341-352.

Webster, C. J. (1996). Urban morphology fingerprints. *Environment and Planning B* **23**, 279-297.

Weng, Q. H. (2002). Land use change analysis in the Zhujiang Delta of China using satellite remote sensing, GIS and stochastic modelling. *Journal of Environmental Management* **64**, 273-284.

Wu, Q., Li, H. Q., Wang, R. S., Paulussen, J., He, Y., Wang, M., Wang, B. H., and Wang, Z. (2006). Monitoring and predicting land use change in Beijing using remote sensing and GIS *Landscape and Urban Planning* **78**, 322-333.

Xia, Z., and Henderson, F. M. (1997). Understanding the relationships between radar response patterns and the bio- and geophysical parameters of urban areas. *IEEE Transactions on Geoscience and Remote Sensing* **35**, 93-101.

Yang, L., Xian, G., Klaver, J. M., and Deal, B. (2003). Urban land-cover change detection through sub-pixel imperviousness mapping using remotely sensed data. *Photogrammetric Engineering and Remote Sensing* **69**, 1003-1010.

Yu, S., Berthod, M., and Giraudon, G. (1999). Toward robust analysis of satellite images using map information - Application to urban area detection. *IEEE Transactions on Geoscience and Remote Sensing* **37**, 1925-1939.

Zha, Y., Gao, J., and Ni, S. (2003). Use of normalized difference built-up index in automatically mapping urban areas from TM imagery. *International Journal of Remote Sensing* **24**, 583-594.

Zhang, J., and Foody, G. M. (1998). A fuzzy classification of sub-urban land cover from remotely sensed imagery. *International Journal of Remote Sensing* **19**, 2721-2738.

Zhang, Y. (2001). Detection of urban housing development by fusing multisensor satellite data and performing spatial feature post-classification. *International Journal of Remote Sensing* **22**, 3339-3355.

INDEX

A

AAS, 38, 150
abatement, 187
abiotic, viii, 47, 70
absorption, 26
academic, 126
access, 61, 89, 113, 193
accessibility, 194, 199
accounting, 105
acculturation, 38
accuracy, 107, 129, 168, 195, 196, 202, 206
acid, 42, 44, 179, 180
ACS, 38
acute, 49, 52
ad hoc, 136
adjustment, 8, 13, 107
administration, vii, 3, 132, 140
administrative, 5, 125, 126, 132, 139, 140
administrators, 41
adult, viii, 37, 47, 50, 51, 52, 60, 61, 68, 73, 105, 106, 111, 119, 150, 151, 198
adulthood, 39
adults, 38, 53, 54, 106
Aedes, 49, 53, 58, 59, 61, 62, 63, 65, 67, 69, 72, 74, 75, 76, 77, 78, 79, 81
aetiology, 106, 107, 108, 119
Afghanistan, 127
AFM, 38
Africa, vii, viii, xi, 35, 36, 37, 39, 40, 41, 42, 43, 49, 52, 57, 58, 60, 61, 62, 76, 82, 84, 127, 191, 192, 193, 199, 203, 204, 205, 206
African Americans, 41
age, 38, 39, 43, 48, 84, 107, 108, 110, 111, 112, 113, 114, 117, 118, 128
ageing (aging), 111, 112, 113, 122
agenda-setting, 136
agent, 50, 54
agents, 56, 88, 95, 96, 100, 101, 109, 129
aggregation, 115, 116, 180
agricultural, x, 6, 51, 57, 58, 64, 67, 68, 76, 77, 86, 108, 109, 110, 122, 126, 128, 129, 141, 145, 161, 162, 173, 174, 180, 182, 206
agricultural sector, 126
agriculture, vii, 6, 50, 51, 57, 58, 60, 63, 71, 108, 171, 195
aid, xi, 93, 191
aiding, 194
AIDS, 200
air, 17, 18, 21, 22, 25, 27, 29, 52, 55, 60, 66, 67, 178, 179
air travel, 67
airports, 71
Albania, 141
alcohol, viii, 35, 41
alcohol abuse, viii, 35
algae, 53, 148, 150
algorithm, 27, 201
alkali, 7
alluvial, 164
alpha, 148
ALS, ix, 105, 106, 107, 108, 109, 110, 111, 112, 113, 114, 115, 116, 117, 118, 119, 120, 121, 122
alternative, x, 161, 174, 195
alternatives, 41, 75, 162
alters, viii, 47, 59, 60, 76
Amazon, 81
ambient air, 52, 60
ambient air temperature, 52, 60
amendments, 125
ammonium, 147
amphibians, 151, 154, 158
amyotrophic lateral sclerosis, 119, 120, 121, 122
analytical framework, ix, 125
Angola, 127
animals, 57, 144, 149
ANN, 201
Anopheles gambiae, 75, 80
Anopheles mosquitoes, 53
anthropic, x, 143
anthropic factors, x, 143
anthropogenic, viii, 48, 50, 64, 68, 143, 146, 147, 148, 149, 150, 158

antibodies, 78
anxiety, 37
appendix, 99, 101
application, 4, 13, 136, 193, 194, 195, 196, 204
aquatic systems, 75
Aquifer, 172, 174
aquifers, x, 161, 162, 163, 164, 166, 170, 171, 172, 173, 174, 175
Arboviruses, 79
argument, 89, 94
arid, 206
arsenic, 148
arteries, 39
arthropods, 77
Asia, 49, 52, 57, 61, 84, 127, 140, 141, 162, 192
assumptions, 27, 84, 90, 196
atherosclerosis, 39
Atlas, 199, 200, 203
atmosphere, 165, 193
Australia, 58, 59, 61, 63, 65, 66, 67, 68, 74, 75, 76, 78, 79, 80, 81, 127, 206
Austria, 140
authority, 125
availability, x, 56, 113, 161, 171, 174
averaging, x, 161
awareness, 67, 111, 113

B

bacteria, 53
bacterial, 82
bacterium, 50, 61
Bali, 186
Bangladesh, 127
barriers, 62, 141, 200
baths, 68
beaches, 7
behavior, 27, 42, 51, 52, 54, 56, 58, 65, 75, 77, 88, 101, 179
Beijing, 3, 13, 126, 130, 131, 139, 140, 141, 189, 207
benefits, 44, 55, 85, 91, 113, 173, 189, 199
Bhutan, 127
bias, 113, 168
biodiversity, viii, 47, 56, 58, 61, 62, 64, 73, 74, 78, 79, 145, 158
biomass, 149
biosecurity, 66, 71
biota, 50, 58
biotic, viii, 47, 59, 62, 70
biotic factor, 70
birds, 68
birth, 40, 108, 110, 111
birth rate, 111
birth weight, 40
black women, 41, 43
blood, x, 38, 39, 40, 41, 42, 43, 44, 53, 68, 77, 143, 145, 151, 153, 154, 158
blood pressure, 38, 39, 40, 41, 42, 43, 44
blood vessels, 39
BMI, 38, 39
bogs, 147, 149
borderline, 110, 113, 117
boreholes, 171
bottom-up, 135
boundary conditions, 18, 28
Boussinesq, 25
brainstem, 105
Brazil, 44, 55, 59, 60, 80, 199
breeder, 65, 66, 67
breeding, 6, 51, 52, 56, 57, 59, 60, 61, 63, 65, 66, 68, 70, 73, 74, 79, 81, 147, 198
buildings, vii, 17, 22, 24, 28, 30, 31, 32, 33, 58, 60, 162
Burkina Faso, 127
burning, 29, 30, 31, 32, 34
Burundi, 127
butane, 150

C

cabbage, 42
cadmium, 150, 151, 154, 156
calcium, 144
calibration, 162, 166, 167, 168
Cambodia, 127
Cameroon, 72
Canada, 150
capacity, 50, 53, 54, 146, 162, 168, 172, 173, 174
capital accumulation, 86, 89, 93, 94, 95, 97
capital markets, 88
capitalism, 136, 140
capitalist, 138
carbohydrate, 37
carbon, 25
carbon dioxide, 25
cardiovascular disease, viii, 35, 41, 42, 43
cardiovascular function, 43
Caribbean, 127
carrier, 65, 179
case study, viii, x, 47, 177, 201
categorization, 107
cats, 64
causation, 108, 118
cell, 36
cellular automaton, 201
Centers for Disease Control (CDC), 37, 53, 72
central city, 125
centralized, 135, 146
certificate, 130
certification, 106
CFD, vii, 17, 18, 20, 25, 27, 32, 33
Chad, 127
changing population, 121
channels, 124
chemical agents, x, 143
chemical properties, 56
chemical reactions, 25

chemicals, 108, 109, 110, 120
childhood, 39, 40, 87
childrearing, 88
children, viii, 35, 37, 39, 40, 41, 43, 84, 85, 86, 87, 88, 89, 90, 91, 92, 93, 94, 95, 97, 98, 99, 100, 101
Chile, 78
China, ix, 3, 13, 32, 55, 123, 124, 125, 126, 127, 128, 129, 130, 131, 132, 133, 134, 135, 136, 137, 138, 139, 140, 141, 142, 177, 179, 189, 203, 207
chloride, 170
cholesterol, 37, 38
chronic diseases, viii, 35, 36, 40
chronic stress, 40
Cincinnati, 81
circulation, 132, 144
citizens, 146
citizenship, 133, 137, 138, 141
cladocerans, 150
classes, 195
classification, vii, 3, 4, 5, 6, 7, 8, 9, 10, 12, 79, 81, 107, 176, 195, 196, 197, 200, 201, 202, 203, 204, 205, 206, 207
clay, 162, 164, 165, 166
climate change, viii, 48, 67, 74, 77
clinical diagnosis, 106
closure, 18
clustering, 124
clusters, 63, 108
CO2, 25, 67
coastal areas, 129
Coca-Cola, 36
collectivism, 38
combustion, vii, 17, 18, 24, 25, 26, 27, 32, 33
commerce, 56
commodity, 194
communication, 172, 178
communities, 41, 42, 61, 62, 70, 72, 74, 137, 144, 145, 149
community, 37, 42, 52, 58, 61, 62, 68, 76, 108, 114, 188
compaction, 162
competence, 52, 60, 65, 81
competition, 53, 61, 62, 63, 68, 70, 72, 73, 75, 79, 112, 122, 126
competitor, 62, 153
complement, 158
complementarity, 87
complex interactions, 138
complexity, 27, 137, 195
compliance, 40
components, 69, 128, 144, 159, 174
composites, 194
composition, x, 60, 76, 112, 113, 117, 138, 143, 145, 146, 148, 150, 152, 153, 193, 197, 205
compounds, 144, 145, 148
compressibility, 25
computation, 127
Computational Fluid Dynamics (CFD), 17, 19, 21, 23, 25, 27, 29, 31, 33
computational grid, 18, 20
computing, 199
concentration, 20, 25, 84, 118, 124, 129, 132, 148, 154, 177, 178, 189
conception, 179
conceptual model, 135
conceptualization, 134
concrete, 13, 20
condensation, 186
conditioning, 18, 21, 22
conductance, 168
conduction, vii, 17, 18, 20, 22, 32, 33
conductivity, 20, 164, 165, 166
confidence, 107, 110, 117
confidence interval, 107, 110, 117
conflict, 57, 141
conjecture, 136
conservation, 182
construction, ix, 5, 12, 53, 60, 125, 173, 175, 181
consumers, 147, 175
consumption, 42, 58, 87, 88, 135, 172
consumption patterns, 42
contamination, 170
continuity, 18
control, viii, 33, 48, 49, 68, 69, 70, 75, 79, 80, 108, 110, 120, 137, 148, 154, 162, 174, 179, 181, 197, 202, 203, 204, 205
convection, vii, 17, 18, 20, 24, 32, 33, 34
convective, 18
convergence, 71, 132
conversion, 25, 58, 180
cooling, 22, 24, 32, 33, 150
copepods, 150
coping, 38
coping strategies, 38, 41
copper, 150, 154, 155, 157
coronary heart disease, 44
correlation, 12, 115, 120, 124
correlations, 38
cortisol, 40, 43
Costa Rica, 102
cost-effective, viii, xi, 35, 40, 71, 191
costs, x, 55, 85, 88, 129, 135, 161, 172, 173, 199
countermeasures, 188
coupling, 20, 183, 201
coverage, 20, 167, 199, 200
covering, 33, 64
CRC, 72, 119
creativity, 71
crime, 41
crops, 57, 175
cross-country, 85, 97
cross-sectional study, 122
crustaceans, 154
cues, 73
cultivation, 61, 175
cultural factors, 179
culture, 7, 41, 188
Cybernetics, 203

cycles, 51, 57

D

data set, 205
database, 202
DBP, 38, 39
death, 52, 108, 110, 111
death rate, 111
deaths, 53, 108, 120
decay, x, 25, 177, 180, 185, 187
decentralization, 124, 136
deficiency, 147, 148, 154, 158
deficit, x, 143
definition, ix, 3, 105, 110, 115, 116, 117, 118, 126, 127, 139
deforestation, 50, 60, 62, 63, 64, 81
degradation, 81, 147
delivery, 40, 172
demand, x, 161, 162, 174
dementia, 105
demographic characteristics, 83
demographic data, 112
demographic factors, 108
demographic transition, 84, 85
demographics, 53
demography, 119
dengue, 52, 59, 61, 62, 63, 65, 66, 67, 71, 73, 74, 75, 77, 79, 81, 193
dengue fever, 73, 193
dengue hemorrhagic fever, 74
density, viii, 25, 26, 47, 50, 55, 57, 59, 61, 68, 89, 109, 178, 179, 181, 198, 204
dependent variable, 135
deposition, 181
deposits, 145, 150, 159
depression, 37, 187
desiccation, 53, 56
desire, 87, 123
destruction, 68, 181
detection, 41, 77, 194, 196, 197, 202, 203, 204, 205, 207
detoxification, 109
detritus, 53, 82
developed countries, 84, 98, 127
developed nations, 48, 55
developing countries, 36, 78, 84, 88, 98, 203, 204
developing nations, 48, 51, 55, 57, 59
development policy, 125
diabetes, viii, 35, 41
diagnostic criteria, 106, 107, 113
Diamond, 68, 74
diarrhea, 40
diastolic blood pressure, 39
diastolic pressure, 39
diatoms, 148, 150
diet, viii, 35, 36, 37, 42
dietary, 39, 42
diffusion, 20, 175
disaster, 185
disease model, ix, 48, 71
diseases, viii, ix, 35, 39, 40, 41, 42, 47, 48, 49, 50, 51, 52, 53, 54, 55, 56, 57, 58, 59, 62, 63, 64, 66, 67, 70, 71, 73, 74, 78, 106, 108, 112, 119, 121, 193, 204
disorder, 105
dispersion, 177, 178, 180, 183, 198
displacement, 63
dissolved oxygen, 147
distribution, 18, 22, 23, 24, 28, 30, 33, 36, 37, 49, 54, 58, 59, 61, 67, 70, 73, 77, 79, 87, 89, 93, 106, 109, 116, 119, 120, 143, 168, 178, 179, 180, 182, 183, 185, 193, 197, 201, 206
district heating, 33
diversity, 11, 61, 74, 76, 106, 143, 147, 149, 154, 159, 198
division, vii, 3, 158
dominance, 11, 70
drainage, 58, 60, 144, 181, 182
drinking, 57, 60
droughts, 173
dumping, 68
duration, 124
dust, 178

E

earnings, 84
earth, 13, 58, 181, 182, 193, 205
eating, viii, 35, 42
Ebola, 49
ecological, viii, ix, 12, 47, 48, 49, 50, 52, 55, 56, 58, 59, 60, 62, 64, 69, 71, 81, 82, 144, 145, 147, 148, 158, 159, 189
ecologists, 61
ecology, ix, 12, 13, 49, 53, 69, 71, 72, 73, 76, 77, 78, 79, 82, 143, 158, 159, 189
economic development, ix, 83, 84, 85, 86, 87, 97, 98, 99, 127, 137, 140, 189
economic growth, 84, 86, 132, 137, 162
economic status, 41, 71
ecosystem, viii, ix, 12, 47, 48, 50, 59, 64, 71, 72, 74, 78, 81, 144, 145
ecosystems, ix, x, 48, 58, 64, 71, 72, 82, 143, 144, 145, 150, 157, 158, 159, 201
education, 6, 13, 102, 186
egg, 53, 63
Egypt, 175
elaboration, 136
elderly, 87, 109, 111, 112, 114
elderly population, 114
electromagnetic, 193
electromyography, 113
emission, 20, 26, 187, 188, 194
emission source, 194
emotional, 37
employees, 129, 130
employers, 130

employment, 51, 57, 108, 128, 129, 130, 137, 139, 189
encephalitis, 49, 53, 57, 62, 65, 73, 77, 81
endocrine, 40, 41
endothelial dysfunction, 39
energy, vii, viii, 18, 19, 20, 26, 35, 36, 37, 58, 60, 82, 173, 175
engines, 84
England, 65, 98, 108, 121, 122
enlargement, 133, 177
entrepreneurs, 124
environment, vii, viii, ix, x, 12, 17, 18, 20, 24, 32, 33, 34, 47, 48, 54, 55, 57, 60, 63, 64, 65, 77, 79, 80, 105, 109, 110, 114, 115, 117, 118, 119, 122, 136, 144, 149, 150, 154, 159, 161, 162, 177, 178, 179, 180, 181, 182, 185, 188, 189, 198, 201, 204, 205
environmental change, ix, 47, 50, 53, 55, 58, 64, 68, 72, 74, 105, 118
environmental conditions, 50
environmental effects, 175
environmental factors, ix, 53, 70, 105, 106, 107, 108, 111, 112, 114, 120, 159
environmental impact, 58, 60
environmental protection, 79
epidemic, 50, 56, 62, 65, 108, 120
epidemics, 40, 61, 73, 75
epidemiology, 55, 70, 74, 78, 79, 106, 110, 119, 120, 122, 197, 205, 206
Equatorial Guinea, 127
equilibrium, 85, 86, 87, 88, 90, 91, 93, 94, 95, 96, 97, 98, 99, 101, 147
Eritrea, 127
erosion, 60, 182
ESR, 64, 74
estimating, 202
estuarine, 58
Ethiopia, 127, 200
ethnicity, 43
ETI, 175
etiology, 121
Europe, 55, 58, 84, 98, 102, 103, 104, 106, 110, 127, 149, 192
European Space Agency, 204
Europeans, 63
evaporation, 18
evapotranspiration, 164, 165, 166, 169
evolution, 72, 77, 131, 135, 145, 177, 179, 180
exclusion, 42, 75
exercise, 41, 173
expansions, 56, 139
expert, 204, 206
expert systems, 204
exploitation, 162, 170, 172, 173, 174, 180, 181, 185
exposure, ix, 37, 48, 61, 66, 70, 71, 107, 108, 109, 110, 120, 198
extinction, 63, 158
extraction, x, 147, 161, 171, 172, 173, 174, 195, 203, 205

F

failure, 63, 106, 114
falciparum malaria, 81, 206
familial, 105
family, 36, 38, 41, 53, 109, 151
family history, 109
family members, 36
FAO, 42
farmers, 9, 42
farming, 64, 110, 128
farmland, 124, 182, 188
farmlands, x, 161
farms, 108, 115
fasting, 38
fat, 36, 37
fats, viii, 35
fatty acids, 36, 42
fauna, viii, ix, x, 48, 62, 64, 65, 67, 68, 73, 143, 145, 146, 147, 149, 150, 152, 153, 157, 158, 159
FDI, 129, 132
fear, 37
fee, 53
feedback, 40
feedback inhibition, 40
feeding, x, 39, 53, 54, 56, 58, 75, 143, 144, 182, 185
females, 39, 40, 53, 68, 108
fertility, ix, 83, 84, 85, 86, 87, 90, 91, 92, 93, 97, 98, 99
fertility rate, ix, 83, 84, 85, 87, 90, 91, 92, 97, 98, 99
fetus, 40
fever, 49, 52, 80
fiber, viii, 35
filariasis, 49, 52, 57, 65, 67, 73, 79
filters, 151
finance, 7
financial resources, 41
financial support, 99
finite volume method, 26
Finland, 109, 121
fire, vii, 17, 18, 24, 25, 27, 28, 29, 30, 31, 32, 33, 34
fires, 24, 33, 63, 194
firms, 89, 129, 178, 179, 184, 186, 187, 189
fish, x, 143, 145, 146, 147, 151, 154
fisheries, 63
fishing, 58, 146
flame, 150
flaviviruses, 77, 80
flight, 25, 27, 68
float, 53
floating, 124, 130, 139, 140
flood, 71, 181
flooding, 162
flora, 64, 146
flora and fauna, 64, 146
flow, vii, x, 17, 22, 25, 27, 28, 29, 30, 31, 32, 82, 129, 133, 161, 162, 165, 166, 169, 173, 174, 175, 188

flow field, vii, 17, 32
flow rate, 29, 173
fluctuations, 149, 171, 173
fluid, 25, 27, 33, 137
focusing, 50, 135
folate, 42, 44
folic acid, 42
food, x, 40, 41, 42, 44, 77, 143, 144, 145, 153, 154, 158
foreign direct investment, 128
forest fragments, 67
forestry, 13, 195
forests, 62, 64, 67, 68, 71, 73, 149
fortification, 39
fossil, 171
fragmentation, 11, 12, 206
France, 98, 121, 122, 205
fresh water, 144
freshwater, ix, 143, 145, 151, 159, 170, 198
friction, 179, 180
fuel, 25, 26
fungi, 149
fusion, 195

G

garbage, 178
gas, 26, 29, 33, 184
gas phase, 26
gases, 188
Gaussian, 28
GDP, 126, 128, 132, 137
generalization, 4, 136
generation, 22, 29, 30, 34, 136
genetic traits, 109
genetics, 109
Geneva, 42, 43, 81, 82
geochemical, 171
geography, 158
geology, 166, 179
geophysical, 207
girls, 39
GIS, xi, 191, 193, 195, 196, 199, 201, 203, 205, 207
glaciers, 7
global climate change, 67
global economy, 66
globalization, 50, 56, 66, 77, 124, 125, 136, 138, 140, 141, 204
goals, x, 71, 177, 187, 197
google, 193
governance, 126, 133, 139
government, 7, 41, 48, 64, 107, 124, 129, 132, 137, 138
grain, 6, 171
granites, 146, 147
grass, 6, 7, 10
grassland, 8
gravity, 27, 172
Great Leap Forward, 125
green land, 5, 7, 10, 12
greening, 34
groundwater, x, 57, 161, 162, 163, 164, 165, 166, 167, 168, 169, 170, 171, 172, 173, 174, 175, 179, 182, 185, 186, 187
groups, x, 43, 86, 87, 114, 143, 145, 147, 149, 150, 154
growth, ix, x, 33, 48, 55, 58, 60, 69, 72, 83, 84, 85, 86, 95, 97, 98, 123, 124, 125, 128, 129, 131, 135, 136, 137, 139, 140, 141, 142, 161, 162, 171, 174, 192, 197, 201, 203, 205, 207
growth rate, 60
Guam, 109
Guangzhou, 132, 134, 141
Guinea, 74, 127

H

habitat, x, 51, 57, 58, 59, 63, 66, 68, 69, 70, 71, 75, 143
Haiti, 127
handling, 136
hanging, 140
harm, 180
harmony, 180
Hawaii, 140
health, viii, ix, xi, 35, 36, 37, 39, 40, 41, 42, 44, 48, 50, 51, 52, 55, 57, 66, 71, 72, 73, 74, 75, 77, 78, 79, 81, 107, 111, 114, 144, 159, 191, 193, 202, 203, 204, 205
health care, 55, 193
health effects, 36, 57
health information, 41
health problems, 40
health services, 52, 111, 114
healthcare, 194, 200
heart, 44, 112
heart disease, 44, 112
heat, 17, 18, 20, 21, 22, 24, 25, 29, 32, 52, 58, 60, 164
heat release, 21, 22, 29, 32
heat transfer, 18
heating, 33, 118, 150
heavy metal, x, 143, 144, 145, 146, 150, 154, 158, 159
heavy metals, x, 143, 144, 145, 146, 150, 154, 158, 159
Hedonism, 38
height, 20, 22, 24, 146
helplessness, 37
heredity, 109
heritability, 120
heterogeneity, 39, 193, 196, 201
heterogeneous, 135, 195, 196
high blood pressure, 39, 40, 41
high resolution, 202, 205
high tech, 158
higher quality, 199
hirudofauna, x, 143

histogram, 168
HIV, viii, 35, 40, 52, 193, 200
HIV/AIDS, viii, 35, 40, 52
Holland, 202
Hong Kong, 123, 139, 141
hopelessness, 37
Hopfield neural network, 206
hormone, 107
hospitalization, 52
host, 49, 50, 52, 53, 54, 55, 56, 57, 61, 73, 76
host population, 55, 56
House, 103, 141, 158, 159
household, 124
households, 36, 63
housing, 7, 9, 10, 37, 142, 207
HPA, 40
HPA axis, 40
human, vii, viii, ix, 7, 47, 48, 49, 50, 51, 52, 53, 54, 55, 56, 57, 58, 59, 60, 61, 63, 64, 65, 66, 68, 69, 71, 72, 73, 74, 75, 77, 79, 81, 83, 84, 85, 86, 87, 89, 90, 91, 92, 93, 94, 95, 96, 97, 99, 100, 101, 114, 118, 179, 181, 182, 193, 194, 198, 203, 204
human animal, 57
human capital, ix, 83, 84, 85, 86, 87, 89, 90, 91, 92, 93, 94, 95, 96, 97, 99, 100, 101
human immunodeficiency virus, 193, 204
humanity, 48
humans, 40, 48, 50, 51, 54, 56, 57, 58, 61, 63, 64, 65, 71, 193
humidity, vii, 17, 18, 21, 23, 24, 32
hunting, 63
hybrid, 21, 201
hydrates, 36
hydrobionts, x, 143, 145, 147, 149, 150, 154, 158
hydrochemical, 171
hydrogeology, 170
hydrologic, 165, 174
hydrological, 149
hydrology, viii, 5, 47, 59
hypertension, viii, 35, 37, 38, 39, 40, 41, 43, 44
hypothesis, ix, 39, 105, 110, 112, 140, 141, 171

I

IARC, 122
ice, 172
identification, 77, 108, 194, 195, 206
ideology, 124
idiopathic, 105, 118
idiosyncratic, 137
Illinois, 47
image analysis, 195
imagery, 193, 194, 195, 196, 197, 199, 200, 201, 202, 203, 204, 205, 206, 207
images, 193, 197, 201, 202, 204, 205, 206, 207
imaging, 194, 195, 206
immigration, 64
immunity, 51, 56, 57
implementation, x, 138, 162
in transition, 43, 137, 141, 203
in utero, 40
inactive, viii, 35, 37
inauguration, 124
incentive, 84
incidence, viii, ix, 35, 39, 52, 55, 56, 57, 58, 61, 64, 105, 106, 107, 108, 109, 110, 111, 112, 113, 114, 115, 116, 117, 118, 119, 121, 122
incineration, 150
inclusion, 133, 196
income, ix, xi, 40, 83, 85, 86, 87, 88, 97, 98, 99, 126, 127, 191, 192, 193, 194, 196, 199, 200, 205, 206
income distribution, 87
income inequality, ix, 83, 85, 86, 87, 97, 98, 99
incompressible, 25
incubation, 54, 60
incubation time, 54, 60
independence, 194
independent variable, 135
India, 55, 73, 175, 199
Indian, 73
indicators, 41
indices, 195, 198
Indigenous, 42, 44
individualism, 38
induction, 109
induction period, 109
industrial, x, 58, 62, 98, 114, 124, 132, 135, 145, 146, 148, 154, 158, 161, 162, 163, 166, 174, 175, 178, 179, 182, 184, 185, 186, 188, 189
industrial production, x, 161
industrial revolution, 98
industrialization, vii, 64, 108, 124, 128, 129, 135, 139, 140, 181, 189, 196
industrialized countries, 123
industry, vii, 5, 7, 48, 144, 169, 178, 183, 188, 189
inequality, 85, 86, 87, 94, 96, 97
infant mortality rate, 97
infection, 50, 52, 57, 58, 59, 61, 62, 64, 65, 66, 69, 73, 79, 80, 198, 205
infections, 40, 52, 57, 64, 66
infectious, viii, 35, 40, 47, 49, 54, 55, 56, 57, 72, 73, 74, 76, 77, 78, 82
infectious disease, viii, 35, 40, 47, 55, 57, 72, 73, 74, 76, 77, 78, 82
inflation, 125
information systems, xi, 191, 193
infrared, 193, 194, 201
infrastructure, 55, 67, 89, 132, 181, 187, 203
inheritance, 106
inhibition, 40
initiation, 125
injuries, 120
innovation, 71, 89
Innovation, 102, 203
insecticides, 68
insects, 63, 66, 72, 74, 147, 151, 154
inspection, 5
instability, 168

institutions, 124, 138
insulin, 38
insurance, 7
integration, 139, 203
integrity, 181
intensity, 26, 55, 89, 93, 178, 179, 182
interaction, x, 4, 59, 114, 132, 177, 178, 179, 180, 181, 182, 185
interaction process, 178
interactions, 52, 58, 61, 63, 70, 80, 138
interdisciplinary, 79, 82
Intergovernmental Panel on Climate Change (IPCC), 67, 78
interpretation, 99, 196
interval, 28, 29, 54
intervention, 70, 81, 194, 198
intervention strategies, 194
intrinsic, ix, 48, 187
invasive, viii, 47, 50, 56, 75, 77
invasive species, viii, 47, 50
invertebrates, x, 67, 143, 147, 149, 151, 152, 154, 159
investigations, 70, 159, 175
investment, 86, 89, 90, 91, 128, 135, 141, 172, 173, 179, 180
investors, 138
Ireland, 109, 119, 121
irrigation, 76, 166, 173, 174, 175
island, 8, 9, 17, 52, 58, 60
isolation, 63, 65, 66, 68, 74, 124, 136
isotope, 176
Israel, 109, 121, 201
Italy, ix, 105, 108, 109, 110, 118, 119, 120, 121, 122

J

Japan, x, 17, 20, 34, 63, 65, 83, 99, 110, 122, 127, 161, 162, 163, 164, 165, 166, 167, 169, 171, 172, 173, 175, 176
Japanese, 33, 34, 49, 57, 65, 76, 77, 81, 175, 176
Japanese encephalitis, 49, 65, 77
Java, 170
jobs, 138
Jun, 189
jurisdiction, 132
justification, 58

K

Kazakhstan, 144
Kenya, 191, 205, 206
kinetic energy, 20
King, 43, 73, 74
Kiribati, 127
Kuwait, 204

L

labor, 88, 129, 130, 136, 178
labor-intensive, 178
Lagrangian, 27
lakes, x, 143, 144, 145, 148, 149, 153, 154, 159
land, vii, 3, 4, 5, 6, 7, 8, 9, 10, 11, 12, 13, 51, 56, 57, 58, 64, 67, 76, 77, 78, 124, 128, 132, 133, 137, 140, 141, 142, 148, 162, 165, 166, 171, 175, 176, 177, 178, 180, 181, 182, 188, 189, 193, 194, 195, 196, 197, 199, 200, 201, 202, 203, 204, 205, 206, 207
land use, vii, 3, 4, 5, 6, 7, 8, 11, 12, 67, 76, 77, 78, 132, 166, 177, 180, 181, 189, 195, 196, 197, 200, 201, 203, 204, 205, 206, 207
Land Use Policy, 201
landscapes, x, 48, 51, 53, 57, 58, 59, 60, 62, 63, 64, 67, 71, 78, 138, 140, 143, 145, 152, 154, 158, 201
land-use, 3, 5, 11, 58, 124, 203
large-scale, 5, 56, 64, 67, 178, 181
larvae, 53, 60, 69, 75, 79
larval, viii, 47, 50, 53, 60, 61, 63, 66, 68, 69, 70, 71, 72, 76, 77, 147, 149, 151, 154
later life, viii, 35, 37, 40
lateral sclerosis, 119, 120, 121, 122
Latin America, 127, 192
law, 21, 28
laws, 179
LDL, 38
leach, x, 143
leaches, x, 143
lead, viii, 35, 37, 38, 50, 52, 58, 60, 62, 70, 71, 92, 150, 151, 154, 156, 170, 174, 195
leadership, 124
legislation, x, 162
leisure, 12
Liberia, 127
life expectancy, 113, 120, 122
life style, 179
lifecycle, 50, 53, 54
lifestyle, viii, 35, 36, 40, 42, 112
likelihood, viii, 48, 50, 56, 60, 128, 196, 204, 205
limitations, 106, 107
linear, 87, 178, 179, 183, 185
links, 58, 79
lipid, 39
lipid profile, 39
lipoprotein, 37
liquor, 36
livestock, 6, 57, 61
living conditions, viii, 35
living standards, 55
local government, 138
location, 20, 172, 178, 179, 187, 193, 194, 197
logging, 188
London, 79, 80, 119, 120, 140, 189, 201, 202, 203, 204, 205, 207
long distance, 178
longevity, 54
long-term, 123, 135, 187, 188, 196, 201
low tech, 90
low-density, 37
low-density lipoprotein, 37

low-income, xi, 191, 192, 193, 200, 205
lying, 63, 132
lymphatic, 57, 65

M

machines, 53, 196, 201
macronutrients, 36
mainstream, 136
maintenance, 173
major cities, 133
malaria, xi, 49, 52, 53, 57, 60, 61, 65, 67, 71, 72, 73, 74, 76, 79, 80, 191, 193, 197, 198, 199, 202, 203, 204, 206
males, 36, 39, 40, 43, 107, 108
malnutrition, viii, 35, 40, 57
mammals, 58, 61, 68
management, viii, x, 4, 6, 11, 12, 13, 35, 40, 41, 50, 58, 161, 162, 165, 166, 173, 174, 175, 179, 180, 193, 194, 206
manifold, 64
man-made, 6, 7, 72
manufacturing, 55, 132
Maori, 63, 79
mapping, xi, 191, 193, 194, 195, 199, 200, 203, 205, 206, 207
marginalisation, 141
market, 58, 88, 124, 129, 134, 136, 137, 138, 139, 140, 141, 172, 178
market economy, 124
Markov, 195
Markov random fields, 195
Markovian, 197
Marxist, 135
mask, 199
Massachusetts, 33
Mauritania, 127
meals, 68
mean systolic blood pressure, 37
meanings, 129
measurement, vii, 17, 18, 20, 21, 22, 24, 28, 126, 203
measures, 66, 69, 79, 135, 174, 195, 197, 199
media, 78, 138, 165
medical care, 113
medicine, 73
Mediterranean, 110
Melanesia, 127
men, 37, 38, 43, 107, 108, 111, 112, 113, 114
mesoscopic, x, 177, 187
MET, 37
metabolic, 44
metabolic syndrome, 44
metabolism, 82, 158
metals, 148, 159
meteorological, 24, 164
metropolitan area, 78, 80, 207
Mexican, 120
Mexico, 75, 107
Mexico City, 107
mice, 61
microbial, 60
Micronesia, 127
Microsoft, 193
Middle East, 57
migrant, 129, 132, 136, 137, 138
migrant workers, 129, 138
migrants, 57, 93, 124, 129, 130, 133, 136, 137, 138, 141
migration, ix, 36, 56, 57, 85, 87, 98, 101, 124, 125, 128, 129, 130, 133, 135, 136, 137, 139, 140, 141, 158
military, 7
Millennium, 48, 71, 72, 77, 81
mineralization, 149
mining, 7, 58, 144, 145
minks, x, 143, 145
Minnesota, 109, 111, 113, 119
mirror, 150
misleading, 110, 116
missions, 205
mixing, 170
mixture analysis, 204
MND, 122
mobility, 136, 139
modeling, vii, xi, 17, 18, 25, 32, 33, 165, 175, 191, 203, 207
models, ix, 42, 48, 69, 70, 71, 81, 84, 85, 93, 97, 136, 139, 196, 197, 201, 204, 206
moderate activity, 37
modernization, 109, 123, 181
modules, 5
modulus, 181, 182
moisture, 34, 166
momentum, 18, 27
Monte Carlo, 197
morbidity, 52
morphology, x, 177, 178, 179, 182, 185, 187, 188, 195, 204, 205, 207
mortality, 52, 55, 57, 63, 86, 97, 106, 108, 110, 111, 112, 113, 119, 120, 121, 122
mortality rate, 55, 86, 108
mosaic, 149
Moscow, 158, 159, 204
mosquitoes, viii, 47, 49, 50, 51, 52, 53, 54, 55, 56, 58, 59, 60, 61, 63, 66, 67, 68, 69, 70, 73, 74, 75, 76, 77, 80, 81, 198
motion, 27, 89
motivation, 85, 180
motor neuron disease, 106, 107, 108, 110, 119, 120, 121
motor neurons, 105, 106, 118
mountains, 144, 148, 149
mouse, 158
mouth, 147
movement, 33, 56, 162, 164
Mozambique, 61, 80, 127
multidisciplinary, ix, 48, 71, 199
multiple sclerosis, 114, 121

muscular tissue, 151
Myanmar, 127

N

NASA, 199, 205
nation, viii, 47, 52, 66, 126
national, vii, 3, 5, 7, 41, 98, 107, 109, 110, 124, 125, 129, 132, 199, 200, 206
National Academy of Sciences, 72, 76
native species, 67
natural, viii, x, 3, 6, 7, 8, 48, 50, 54, 57, 58, 62, 63, 67, 68, 77, 128, 143, 144, 145, 148, 152, 153, 154, 157, 158, 159, 165, 170, 174, 179, 180, 181, 182, 185, 187
natural disasters, 185
natural food, 77
natural resources, 63, 174, 180, 187
negative consequences, 174
Nepal, 127
nervous system, 119
Netherlands, 170
network, 167, 174, 189, 204, 205
neural network, 196, 205, 206
neural networks, 196
neurodegenerative, 109
neurodegenerative disease, 109
neurological disease, 112
neurologist, 113, 114
neurons, 105
New England, 80
New South Wales, 74, 78
New York, 13, 33, 43, 52, 72, 73, 81, 82, 103, 104, 106, 119, 121, 122, 141, 142, 206
New Zealand, viii, 47, 48, 50, 63, 64, 65, 66, 67, 68, 69, 70, 72, 73, 74, 75, 76, 77, 78, 79, 80, 81
Ni, 13, 207
Niger, 127
Nigeria, 199, 201
Nile, 49, 52, 53, 61, 62, 64, 65, 72, 74, 75, 77, 80
Nipah, 49
nitrates, 147
nitrogen, 146, 148
nitrogen compounds, 146
Nixon, 206
Nobi Plain, x, 161, 162, 163, 164, 165, 169, 174, 176
noise, 177, 178, 179, 194
non-human, 57, 58
non-human primates, 58
nonlinear, 172
non-native, 59, 62
normal, 56, 144, 147, 150, 168
normal distribution, 168
North America, 49, 55, 61, 62, 63, 75, 80, 106, 192
Northeast, 201
Norway, 122
nuclei, 178, 182, 187
nursing, 114
nursing home, 114
nutrient, 53, 60, 69, 76
nutrients, 40, 53
nutrition, 40
nutritional disorders, 39

O

obese, 41, 44
obesity, viii, 35, 41, 43
obligations, 138
observations, 98, 136, 147, 167, 171
occupational, 107, 120
Oceania, 84, 192
odds ratio, 108, 110
oil, 146, 148, 188
old age, 84, 112
olive, 146
online, 73, 80
open spaces, 198
optical, 193, 194, 195, 200, 202
optimization, 167, 188
organic, 82, 146, 147, 164
organic matter, 82, 164
organism, 159
organizations, 52
orientation, 58
otters, x, 143, 145
overweight, 37, 41
ownership, 5, 130
oxygen, 25, 26, 147, 148, 150

P

P. falciparum, 198
Pacific, 105, 106, 107, 124
pandemic, 59, 74
paper, ix, 5, 18, 81, 83, 84, 86, 87, 88, 90, 99, 104, 124, 125, 126, 139, 141, 151, 175, 184
paradox, 76, 135, 139
paradoxical, 137
parameter, 25, 91
parasite, 50, 67, 151, 154, 198
parasites, x, 52, 62, 143, 145
parasitic diseases, 78
parenting, 84, 88, 89
parents, 84, 85, 87, 88, 90, 91, 92, 94, 97, 99, 100, 101
Paris, 106
PARK, 3
Parkinson, 112
parkinsonism, 105
particles, 26
passive, 145, 193, 194
pathogenesis, 38, 108, 109, 110, 111, 112, 119
pathogenic, 64
pathogenic agents, 64
pathogens, 50, 51, 54, 55, 56, 57, 58, 60, 61, 64, 65, 66, 67, 69, 70, 71, 75, 82
pathophysiology, 43

pathways, 73
patients, 107, 108, 114
pattern recognition, 196, 201
Pb, 151, 156
PDC, 105
Pearl River Delta, 132, 141
peat, 147, 148
pedestrian, 24, 32
pediatric, 52
per capita, 58, 85, 86, 99, 198
per capita income, 85, 86, 99
perception, 135, 138
peri-urban, vii, 35, 36, 48, 61, 67, 69, 71, 132, 204
permeability, 182
personal, 136, 172
Perth, 80
pesticides, 75
phenotypic, 70
phone, 36
photographs, 195
photolysis, 82
physical activity, 37, 41, 107
physical properties, viii, 47, 59, 165, 194
physiological, x, 43, 53, 60, 143, 145
physiology, 159
phytoplankton, 150
piezometric levels, x, 161
pituitary, 40
plankton, 146
planning, 4, 5, 9, 12, 71, 124, 171, 179, 188, 193, 194, 196, 205
plants, 42, 44, 62, 118, 149, 150, 171
plastic, 153
platforms, 202
plausibility, 118
play, x, 83, 128, 143, 200
Pleistocene, 163
Pliocene, 163, 164
poison, 145
poisoning, 145
Poland, 176
polio, 55
pollutant, 20, 179, 180, 188
pollutants, x, 18, 118, 143, 144, 145, 149, 178, 183, 187
pollution, 17, 144, 145, 147, 148, 154, 157, 158, 177, 178, 179, 180, 183, 184, 185, 188, 198
Polynesia, 127
Polynesian, 63
pond, 6, 146, 147, 152
pools, 60
poor, viii, 35, 39, 51, 57, 60, 68, 71, 114, 148, 168, 171
poor health, 51, 57
population, vii, viii, ix, x, xi, 12, 35, 39, 40, 41, 42, 43, 44, 48, 49, 50, 54, 55, 56, 57, 58, 59, 60, 61, 62, 63, 64, 65, 66, 67, 68, 69, 70, 71, 73, 76, 84, 86, 93, 105, 106, 107, 108, 109, 110, 111, 112, 113, 114, 115, 116, 117, 118, 119, 120, 121, 122, 123, 124, 125, 126, 128, 129, 130, 131, 132, 133, 135, 138, 139, 140, 142, 145, 159, 161, 177, 178, 179, 187, 188, 189, 191, 192, 197, 198, 199, 201, 202, 203, 205, 206
population density, x, 48, 50, 57, 59, 71, 109, 110, 161, 198
population growth, 69, 84, 124, 139, 192, 197, 201
population size, 58, 68, 131
porous, 165, 166
porous media, 165
ports, 67, 68, 73
positive correlation, 84
potassium, viii, 35
Potchefstroom, 35
poultry, 6
poverty, ix, 83, 85, 86, 93, 94, 95, 96, 97, 98, 99, 101
poverty trap, ix, 83, 86
power, 20, 136, 137, 138, 194, 195, 200
powers, 138
precipitation, 164
predators, x, 52, 62, 63, 67, 69, 76, 143, 145, 152, 154
prediction, 33, 165, 196, 201
preference, 87
pregnant, viii, 35
pressure, 26, 27, 29, 38, 39, 40, 41, 43, 44, 143, 148, 158, 187, 207
prevention, 41, 197
prices, 173
primary school, 98
priorities, 40, 197, 204
pristine, 67
private, 174
probability, 28, 53
probability density function, 28
production, ix, x, 33, 51, 52, 60, 76, 83, 85, 87, 88, 89, 91, 93, 135, 147, 161, 163, 173, 180, 182, 199
production technology, ix, 83, 85, 87, 91
productivity, vii, 60, 76, 89, 93
program, vii, 5, 17, 18, 32, 64, 66, 165
prolactin, 40
proliferation, ix, 48, 56, 59, 60, 62, 66, 68, 71
promote, viii, 35, 40, 41, 50, 123, 164, 189
propane, 25, 29, 32, 150
proposition, 95, 195
protection, 44, 71, 79, 143, 182
protein, 37
proteins, 53
PSS, 38
psychological variables, 38
psychological well-being, 41
psychosocial stress, 38
psychosocial variables, 38
public, viii, xi, 5, 7, 8, 9, 35, 47, 50, 52, 57, 66, 69, 70, 71, 75, 76, 81, 98, 163, 191, 193, 194, 195, 196, 199, 200, 204, 206
public education, 98

public health, viii, xi, 47, 50, 52, 57, 66, 69, 70, 71, 75, 76, 81, 191, 193, 194, 195, 196, 199, 200, 204, 206
PUFA, 36, 42
pulse, 43
pumping, x, 161, 162, 163, 169, 171, 172, 173, 174, 175
pupal, 53
purification, 145

Q

quality of life, 55

R

race, 109
radar, 194, 195, 197, 202, 206, 207
radiation, vii, 17, 18, 20, 22, 24, 25, 26, 32, 33, 34, 193, 194
radio, 193
rain, 6, 53, 57, 68, 179, 180
rainfall, 59, 67, 164, 165, 166, 168
rainwater, 59, 60
range, xi, 53, 56, 57, 59, 64, 65, 69, 70, 106, 107, 136, 148, 191, 193, 194, 195, 196, 197, 200
RAS, 158, 159
rate of return, 90
rats, 64
raw material, 178
raw materials, 178
reaction rate, 25
reactivity, 37, 43
reality, 27, 135, 136, 137, 138
reception, 36
recognition, 66, 196, 201
reconstruction, 140
recovery, 162
recreation, 7, 146
reduction, 41, 55, 62, 169, 171
reforms, 123, 124, 125, 135, 137, 138, 142
refractory, 69
refugee camps, 57, 195, 201, 202, 205
refugees, 51, 57
regional, 41, 55, 56, 68, 81, 107, 109, 110, 114, 124, 131, 132, 140, 141, 178, 179, 180, 181, 186, 187, 188, 189, 195
registries, 107, 109, 114
regression, 197, 206
regular, 42, 124, 200
regulation, 5, 124, 149
regulations, x, 4, 140, 161, 162, 174
regulators, 62, 68
relationship, ix, 12, 48, 69, 71, 83, 84, 86, 99, 105, 109, 114, 115, 118, 127, 135, 137, 139, 172, 179, 181, 183, 187, 188
relationships, 71, 75, 77, 199, 207
relatives, 124
relaxation, 137
religion, 7
remote sensing, xi, 191, 193, 195, 196, 197, 199, 200, 201, 203, 204, 206, 207
reproduction, 74
Republic of the Congo, 127
research, vii, 7, 8, 69, 70, 79, 81, 89, 108, 114, 118, 119, 122, 134, 136, 140, 145, 151, 158, 165, 195, 196, 197, 199, 206
research and development (R&D), 89
researchers, 84, 91, 126, 138, 158, 193
reservation, 148, 149
reserves, 68, 80
reservoir, 7, 52, 61, 64, 145, 146, 147, 148, 149, 150, 152, 153, 154, 158
reservoirs, x, 57, 143, 144, 145, 146, 148, 152, 153, 154, 155, 156, 158, 159, 175
residential, 5, 58, 67, 68, 124, 129, 132, 178, 180
residuals, 168, 169
resilience, 78
resistance, 38, 62, 72, 120, 145, 158
resolution, 193, 194, 195, 196, 199, 200, 201, 202, 203, 205, 206
resource allocation, 133
resources, x, xi, 5, 41, 48, 56, 61, 63, 68, 126, 135, 136, 144, 158, 159, 162, 173, 174, 175, 176, 178, 179, 180, 181, 187, 188, 189, 191, 193, 200
respiratory, 40, 49, 52
restructuring, 126, 132, 136, 140, 141, 142
retirement, 87, 88
returns, 85, 90
revenue, 180
Reynolds number, 27
rhythm, 180
rice, 57, 61, 173
rice field, 173
Rift Valley fever, 49, 80
risk, vii, viii, 17, 30, 32, 35, 36, 39, 40, 41, 42, 47, 48, 50, 51, 52, 56, 57, 60, 61, 62, 63, 65, 66, 67, 68, 69, 70, 71, 73, 74, 75, 76, 81, 82, 106, 107, 108, 110, 111, 113, 117, 118, 119, 120, 144, 193, 194, 197, 198, 199, 202, 203, 205
risk behaviors, 41
risk factors, viii, 35, 36, 42, 82, 106, 107, 110, 120
risks, xi, 50, 66, 191, 193, 197
river basins, 144
River Po, 183
rivers, x, 63, 143, 144, 145, 146, 149, 150, 163, 168, 169, 173, 179, 180, 183, 184
rodents, 158
roughness, 194
Royal Society, 67, 72, 74, 79, 202, 206
rubber, 6
runoff, 147
rural, vii, viii, ix, xi, 7, 35, 36, 37, 39, 40, 42, 43, 48, 57, 60, 64, 83, 84, 85, 86, 87, 88, 89, 90, 91, 92, 93, 94, 95, 96, 97, 98, 99, 100, 101, 102, 105, 108, 109, 110, 115, 116, 117, 118, 119, 122, 123, 124, 125, 128, 129, 130, 132, 133, 135, 136, 137, 138,

139, 140, 180, 182, 189, 191, 192, 193, 195, 200, 202, 204, 206
rural areas, xi, 36, 48, 60, 91, 92, 108, 124, 129, 130, 132, 191, 195, 204
rural communities, 39
rural people, 36, 57, 87, 98
rural population, 48, 108, 115, 116, 123, 125, 133
Russia, 146, 149
Russian, 143, 158, 159, 204
Russian Academy of Sciences, 143
Rwanda, 127

S

safety, vii, 17, 18, 32, 159, 174
saline, 7, 170, 171
salinity, 170, 171, 176
salinization, 163
salt, viii, 35, 41, 162, 170
Samoa, 127
sample, 37, 39, 129
sampling, 67, 68, 150
sand, 7, 146
sanitation, 55, 57, 60, 71
SAR, 201, 202, 203, 205
SARS, 49
satellite, xi, 191, 193, 194, 195, 196, 197, 199, 200, 201, 202, 203, 204, 205, 206, 207
satellite imagery, 195, 196, 199, 202
saturated fat, viii, 35
SBP, 38, 39
scalar, 22
Scandinavia, 106
scatter, 30, 31, 32, 33
scattering, vii, 17, 18, 25, 27, 28, 30, 31, 32, 33
school, 39, 41, 89, 98
sclerosis, 105, 122
sea level, 163
search, 174
seasonal variations, 174
seawater, 170
Secretary-General, 48, 72
security, 40, 129
sedimentation, 60
sediments, 163, 165, 171, 172, 175
seeding, 6, 7
self, 38, 142
self help, 137
semiarid, 206
Senegal, 127
sensing, xi, 191, 193, 195, 196, 197, 199, 200, 201, 203, 204, 206, 207
sensitivity, 37, 169
sensor technology, 200
sensors, 193, 194, 195, 196, 199, 200
septic tank, 60
series, 18, 106, 119, 139, 145, 146, 162, 164, 165, 166, 175, 176, 200
serum, 37
services, 48, 52, 60, 66, 80, 111, 114
settlements, viii, 35, 36, 39, 41, 51, 56, 57, 58, 61, 63, 145, 147, 149, 153, 158, 193, 194, 195, 199, 202
severe acute respiratory syndrome, 49
sewage, 146, 148
sex, 109
shade, 67, 76
Shanghai, 141
shape, 25, 27, 28
shares, 129
sharing, 53, 71, 184
shipping, 56, 66, 67
shores, 146, 147, 148, 149
shortage, 136, 188
sibling, 79
Sierra Leone, 127
sign, 36
signals, 193
signs, 149
similarity, 4
simulation, vii, x, 17, 18, 20, 22, 24, 25, 28, 32, 33, 73, 161, 162, 165, 166, 168, 170, 174, 175
simulations, 197
sites, 51, 52, 56, 59, 60, 61, 68, 74, 75, 79, 146, 198
skin, 151
small mammals, 159
smallpox, 55
smelting, 144
soccer, 41
social behavior, 179
social benefits, 189
social change, 139
social context, 43
social environment, 181
social inequalities, 58
social security, 40
Social Security, 113, 130
social support, 38, 41, 42
socialism, 125, 135, 141
socialist, ix, 123, 124, 125, 127, 128, 129, 130, 131, 133, 134, 135, 136, 137, 138, 139, 141
socioeconomic, 40, 41, 109, 124, 125, 138, 174
socioeconomic background, 109
sodium, 41
software, 199
soil, viii, 3, 5, 47, 53, 59, 66, 166, 179, 181, 182, 188, 189, 205
soil erosion, 179, 181, 182, 188
solar, 18, 20
solid phase, 33
solid waste, 178
Solomon Islands, 127
solutions, xi, 90, 191, 202
solvents, 109, 120
Somalia, 127
somatic symptoms, 37
soot, 25, 26, 33

South Africa, vii, viii, 35, 36, 37, 39, 40, 41, 42, 43, 44
South America, 59, 62
South Pacific, 63, 66
Southeast Asia, 52
Soviet Union, 193
spatial, ix, x, xi, 79, 80, 123, 124, 125, 129, 131, 132, 133, 134, 136, 137, 138, 140, 177, 178, 179, 180, 181, 182, 183, 184, 185, 191, 193, 194, 195, 196, 197, 199, 200, 201, 203, 206, 207
spatial location, 178, 179
spatiotemporal, 203
species, viii, x, 25, 47, 48, 50, 52, 53, 55, 56, 57, 58, 59, 60, 61, 62, 63, 64, 65, 66, 67, 68, 69, 70, 71, 72, 73, 75, 76, 79, 80, 81, 143, 144, 145, 147, 148, 149, 150, 151, 152, 153, 154, 157, 158, 198
spectrophotometric, 154
spectrum, 193
speculation, 136
speed, 123, 126, 179, 180, 181, 182, 187
spheres, 138
spinal cord, 105
sporadic, 105, 106, 109
Sputnik, 193
SSS, 38
St. Louis, 65, 73
St. Louis encephalitis, 65, 73
stability, 145
stabilization, 172
stages, 42, 50, 53, 55, 56, 73, 136, 167, 169, 172
stakeholders, 174
standard deviation, 28
standard model, 25
state control, 130
statistics, 13, 112, 114, 115, 116, 124, 125, 129, 140, 142
steady state, 96, 166, 173
stiffness, 39
stochastic, 207
stochastic model, 207
stochastic modelling, 207
stock, 57, 130, 144
storage, 7, 59, 71, 144, 145, 148, 150, 154, 162, 166, 175
stormwater, 59, 61
strains, 80
strategies, x, 38, 41, 70, 75, 161, 162, 187, 188, 194
stratification, 24, 114, 116, 131
streams, 145, 146, 149, 168
strength, 116, 118
stress, viii, 35, 37, 38, 40, 43
stress-related, 43
stretching, 144
stroke, viii, 35, 40, 41, 112
Sub-Saharan Africa, 44, 193, 203, 205
subsistence, viii, 35, 36, 37
substances, 145, 146, 147
Substation, 186
substrates, 53, 82
suburban, 48, 61, 62, 80, 132, 182, 203
suburbs, vii, 115
Sudan, 127
sugars, viii, 35
summer, x, 33, 144, 146, 147, 148, 150, 161, 173, 174
superiority, 12, 62, 63, 135
supply, viii, x, 35, 36, 40, 42, 98, 129, 130, 146, 158, 161, 162, 163, 172, 173, 174
suppression, 146
surface roughness, 194
surface water, 59, 163, 166, 175, 179, 182
surplus, 128, 129
surveillance, viii, 48, 68, 69, 75
survival, 48, 50, 52, 60, 78, 112, 198
survival rate, 198
surviving, 112
susceptibility, 122
sustainable development, 174, 176
Sweden, 65, 77, 104, 108, 109, 120, 121
Switzerland, 81, 82
sympathetic, 37, 43
symptoms, 37, 49
syndrome, 52
synergistic, 50
synergistic effect, 50
systems, xi, 60, 62, 71, 74, 132, 179, 191, 193, 194
systolic blood pressure, 37

T

T and C, 140
taiga, 149
tangible, 134
tanks, 59
Tanzania, 76, 79, 127
target identification, 206
tariffs, 175
taxa, 50
taxes, 175
technological change, 87, 97
technological progress, ix, 83, 84, 85, 86, 89, 91, 93, 94, 95, 96, 97, 98, 99
technology, ix, 55, 83, 85, 86, 87, 89, 90, 91, 93, 94, 95, 97, 98, 99, 129, 178, 200
television, viii, 35, 36
temperature, vii, 17, 18, 20, 21, 23, 24, 25, 27, 29, 30, 32, 53, 67, 76, 147, 148, 150, 164, 165
temporal, 77, 80, 111, 173, 193, 194, 197, 200, 201, 204, 206
temporal distribution, 77
territory, 144, 145, 146, 147, 149, 153, 157
testosterone, 40, 43
testosterone levels, 40
Texas, 121
textbooks, 170
textile, 185
theory, 55, 78, 137, 196
therapy, 41

threat, x, 50, 71, 161, 174
threatened, 80
threatening, 181
threats, 73
three-dimensional, 18, 165, 175, 193
three-dimensional model, 165
threshold, 60
thresholds, 55, 56
thyroid, 109
ticks, 61
time, viii, 5, 20, 22, 25, 27, 35, 39, 48, 50, 53, 54, 63, 66, 70, 71, 73, 84, 85, 86, 87, 88, 89, 90, 94, 95, 96, 97, 98, 99, 101, 106, 109, 111, 112, 113, 114, 115, 125, 129, 136, 137, 146, 147, 157, 166, 172, 194, 195, 196, 197, 199, 200, 202
time periods, 112, 196
time series, 200
tissue, 151
Togo, 127
Tokyo, 17, 20, 162
tolerance, 53, 75
top-down, 135
topographic, 70
topology, 9
tourism, vii, 3, 9
toxic, 150
toxicological, 151, 157
toxins, 109, 114
tracers, 171
trade, viii, 48, 51, 56, 66, 67, 71, 84, 85, 91, 132, 200
trade-off, 71, 84, 85, 91, 200
traffic, 178
training, 136, 196
traits, 50
trajectory, ix, 125
trans, 102, 169
transfer, 18, 26, 124, 129
transformation, 124, 125, 128, 130, 134, 135, 136, 137, 138, 140, 145, 147, 158
transition, viii, xi, 36, 38, 42, 47, 55, 56, 72, 85, 99, 140, 141, 170, 191, 197, 206
transitional economies, 137
transitions, 192
transmission, xi, 51, 52, 56, 57, 59, 60, 61, 65, 67, 68, 69, 71, 73, 74, 76, 80, 81, 191, 198, 199, 203, 205
transnationalism, 136
transport, viii, 19, 25, 27, 34, 35, 53, 56, 132, 154
transportation, 7, 8, 10, 12, 18, 48, 144, 178, 179, 180
traps, 67, 69
trauma, 40
travel, viii, 48, 51, 56, 66, 67
trees, 6, 7
trend, 109, 110, 111, 113, 121, 124, 125, 128, 137, 178, 183
trial, 124
tribal, vii, 35, 36, 39
triglyceride, 41
trophic chains, x, 143
tuberculosis, viii, 35, 40, 193, 202
turbulence, 21, 25, 27, 28
turbulent, vii, 17, 18, 20, 25, 32, 33
Turbulent, 19, 21
Turkey, 175
Tuvalu, 127
two-way, ix, 48, 50, 71

U

ubiquitous, 178
Uganda, 76, 127
ulna, 148
ultraviolet, 82
uncertainty, 194, 195, 204
UNEP, 58, 81
UNESCO, 176
unification, 189
uniform, 55
United Nations (UN), 40, 48, 49, 55, 58, 72, 81, 109, 115, 123, 127, 141, 192, 206
United States, 49, 63, 72, 80, 81, 102, 106, 201
units of analysis, 196
universities, 89
upload, 82
Urals, x, 143, 144, 145, 148, 149, 151, 152, 153, 154, 155, 156, 157, 158, 159
urban areas, vii, 18, 32, 39, 48, 50, 57, 58, 62, 63, 64, 65, 68, 69, 71, 78, 91, 92, 93, 98, 108, 124, 126, 129, 130, 162, 192, 193, 194, 195, 198, 200, 202, 203, 204, 207
urban centers, x, 49, 65, 67, 132, 161
urban centres, 115, 117
urban population, 12, 58, 64, 108, 109, 115, 116, 123, 124, 125, 126, 128, 131, 135, 138, 140, 142, 178, 179, 205
urban settlement, viii, 35, 56
urbanisation, 206
urbanization, vii, viii, ix, x, xi, 35, 36, 38, 39, 40, 42, 43, 47, 48, 49, 50, 51, 53, 55, 56, 57, 58, 59, 60, 61, 62, 63, 64, 68, 69, 70, 71, 72, 78, 83, 84, 85, 86, 87, 91, 93, 95, 96, 97, 98, 99, 101, 105, 108, 109, 110, 114, 115, 116, 117, 118, 123, 124, 125, 126, 127, 128, 129, 130, 131, 132, 133, 134, 135, 136, 137, 138, 139, 140, 141, 142, 158, 177, 178, 181, 182, 185, 188, 189, 191, 192, 193, 196, 197, 198, 199, 200, 206
urbanized, viii, x, 35, 36, 37, 38, 39, 40, 41, 42, 48, 64, 109, 110, 114, 115, 116, 117, 118, 133, 137, 143, 145, 152, 153, 154, 157, 158
USEPA, 71, 81
USSR, 159, 204
uti, 19

V

values, 12, 21, 22, 107, 112, 149, 154, 164, 165, 166, 167, 168, 173

Vanuatu, 127
vapor, 18, 22, 24
variability, 107
variable, 70, 135, 148, 163, 166
variables, 38, 54, 70, 87, 166, 197
variation, 42, 55, 72, 74, 86, 107
vascular disease, viii, 35, 40
vasoconstriction, 40
vasodilatation, 40
vector, viii, 47, 49, 50, 51, 52, 53, 54, 55, 56, 57, 59, 60, 61, 62, 63, 64, 65, 67, 68, 69, 70, 71, 72, 73, 74, 75, 78, 79, 81, 193, 196, 198, 201, 204
vegetables, viii, 35, 37, 42, 44
vegetation, viii, x, 3, 5, 47, 59, 60, 80, 143, 147, 148, 149, 154, 158, 159, 164, 179, 182, 188, 205
vehicles, 44, 56
velocity, vii, 17, 18, 20, 22, 24, 25, 27, 29, 30, 31, 32, 43, 185, 186, 187, 188
Venezuela, 59, 71
vertebrates, 151
Victoria, 79
village, 9, 128, 132, 147
violence, viii, 35, 37, 40, 41
virus, 49, 50, 57, 58, 61, 64, 65, 66, 74, 75, 77, 78, 79, 80, 81, 193, 204
virus infection, 79, 80
viruses, 49, 64, 65, 74, 77, 81
visible, 193, 194, 201, 204
vortex, 22

W

wage rate, 88
wages, 88
Wales, 74, 78, 108, 121, 122
war, 162
waste water, 60, 183, 184, 188
wastes, 178
water, x, 5, 6, 7, 8, 9, 12, 18, 25, 33, 50, 53, 55, 57, 58, 59, 60, 67, 68, 71, 76, 143, 144, 145, 146, 147, 148, 149, 150, 151, 152, 153, 154, 155, 156, 157, 158, 159, 161, 162, 163, 166, 168, 172, 173, 174, 175, 176, 177, 178, 179, 181, 182, 183, 184, 185, 186, 187, 188, 189
water quality, 76, 147
water resources, x, 162, 173, 176
water supplies, 59
water table, 168, 173
water vapor, 18
waterfowl, 151, 154
watershed, 144, 148, 149, 201
wavelet, 196, 205
weight loss, 41
welfare, 7
well-being, 41, 48, 50, 71
wells, x, 161, 163, 165, 166, 167, 168, 169, 172, 173, 174, 175
West Africa, 204
West Nile virus, 49, 52, 53, 61, 62, 64, 65, 74, 80
western countries, 108, 109, 111, 112
wetlands, 57, 188
wildlife, 71, 72, 73
wind, vii, 17, 18, 20, 22, 23, 24, 27, 28, 29, 30, 31, 32, 34, 68
winter, 144, 147, 148
withdrawal, x, 37, 161, 166, 173, 174, 176
women, viii, 35, 37, 38, 41, 43, 44, 107, 110, 111, 112, 113, 114
wood, 33
woodland, 7, 8
woods, 148
workers, vii, 36, 39, 42, 51, 57, 64, 71, 98, 107, 129, 136, 138
World Bank, 55, 56, 80, 82
World Development Report, 82
World Health Organization (WHO), 2, 36, 37, 42, 43, 52, 74, 82, 204
World Resources Institute (WRI), 52, 58, 64, 82
worms, 145

Y

yellow fever, 58, 63
Yemen, 127
yield, x, 161, 166, 174, 175, 194

Z

zinc (Zn), 150, 151, 154, 155, 157
zoonotic, 65, 78, 82
zooplankton, 147, 150